In Memorium

1893   1974

ARTHRITIS AND RADIOACTIVITY

A STORY OF MONTANA'S FREE ENTERPRISE MINE

Originally known as the
Free Enterprise Uranium Radon Mine

by
WADE V. LEWIS

With Foreword by
PATRICIA LEWIS

Peanut Butter Publishing
Seattle, WA
1994

Second Printing, 1969
Third Printing, Revised and Copyrighted, 1984
Fourth printing, Revised and Copyrighted, 1994
10   9   8   7   6   5
11.0084

Library of Congress Catalog Card Number 64-21092

Printed in the United States of America

Typesetting by Packard Productions

published by
Peanut Butter Publishing
226 Second Avenue West
Seattle, WA 98119
(206) 281-5965

FOREWORD

The Free Enterprise Health Mine has been open to the public and in continuous operation since 1952. The first edition of *Arthritis and Radioactivity* was published in 1955. Nine years later, in 1964, the publication was updated to present important points of study, observation and research determined during those additional years of operation of the Free Enterprise Health Mine.

Due to the passing of time, this present edition reflects only minor updates from the author's original manuscript. This is one man's opinion based on years of observation. It contains the researched facts of his time and has not been altered to reflect modern theories or today's discoveries and technology. The contents of this revised edition preserve the history and integrity of the original.

Patricia Lewis

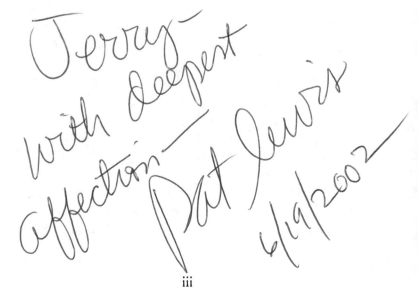

iii

PREFACE

A parallel to the Free Enterprise Health Mine is located near Bad Gastein, Austria, represented by a 1,500 meter adit, radioactive by reason of the radon gas present in its underground mine atmosphere. Medical doctors of Innsbruck University operate this radon adit and nearby spa for the benefit of people afflicted with arthritis and kindred glandular connected ailments. Reports written in German have been received from Austrian doctors, including clinical evaluations tabulated on hundreds of patients. Such reports show the positive benefits received by mine visitors for virtually the same kind of afflictions, and in about the same high percentage of cases as have been observed at the Montana mine.

From 1969 through the 1970s, the company operating the Free Enterprise Health Mine sent representatives to Austria to help further scientific investigation through contact with European doctors on the subject of radon. Due to the dozens of questions which arose following observation of the effect of exposure to radon on arthritics, and other afflicted people, the author was precipitated into research that, of necessity, extended from coast to coast in the United States, and into foreign countries. An endeavor has been made to answer some of these questions.

There are now some 600 radon clinics, medically supervised and government recognized, in Russia, Austria, Germany and other European countries.

Many studies relating to radiation are at present proceeding in the United States under state, federal and private auspices. They cover varied subjects: the many sources of radiation, radiation protection and tolerances, industrial and defense applications of radiation, radiological health, detection of radiation, and peace and wartime uses of atomic energy.

The avenues of research relating to radiation appear limitless, and only a few can be mentioned in a single discussion or text. One scientist could spend a lifetime in study of a single phase of radiation. This manuscript, of necessity, is limited in scope. It is a story of one uranium-radon mine locality in Montana, where radon gas, its transmuted elements, with attending factor of ionization, are found in the Free Enterprise Health Mine workings — such elements now known to have important beneficial effects on afflicted people.

Constructive criticism of the radon subject is most earnestly invited, but many adverse attitudes have been due to lack of knowledge of natural radioisotopes or resistance to enlightenment by refusal to admit the need for study of a new subject. The Radon Research Foundation, a non-profit organization headquartered at Boulder, Montana and terminated in 1976 after the death of Wade V. Lewis in 1974, was a vehicle established for the proper study and required research of the radon subject.

ACKNOWLEDGEMENTS

In preparation of this manuscript the author is indebted to many sources of information newspapers, magazines, periodicals, technical books, doctors, clinics, radiologists, physicists, State and Federal agencies and departments; likewise the Mine visitors themselves for their many letters, documented statements, and personal case histories cheerfully given and faithfully recorded by them. Reference credit to sources of information would have been herein acknowledged in many more instances had not some sources requested their names not be used in publication. The many valuable suggestions and criticisms received from friends during and after manuscript preparation are likewise appreciated and acknowledged with deep gratitude.

CONTENTS

Dedicated

to

Afflicted People Everywhere

CHAPTER 1

ARTHRITIS: ITS CAUSE AND CURE

A *cause* of Arthritis is *known* and Nature *has* a *cure* for Arthritis! This statement is contrary to a popular negative opinion, sponsored by some national organizations purportedly dedicated to aiding people afflicted with arthritis and kindred glandular connected ailments. Such organizations continue to claim that there is no known cause of arthritis and no known cure. Moreover, such organizations continue to solicit and collect millions of dollars from the public each year, "searching for a cure," but refusing to recognize or even investigate radon therapy and the attending factor of ionization supplied by nature and utilized at the Free Enterprise Mine located at Boulder, Montana.

The Free Enterprise Mine, now in its twelfth year of continuous operation, has had many thousands of visitors over this time period. They arrive afflicted with arthritis, bursitis, sinusitis, asthma, eczema or allied ailments. A high percentage of these visitors claim either cure or beneficial response to radon exposure. Many claim complete and permanent relief from the above ailments, while others return after a year or two if threatened by returning symptoms.

Nature's radon therapy effects relief or cure in many cases by reaching a basic cause of rheumatoid or other forms of arthritis, reactivating the body's endocrine gland system toward normal production of ACTH (corticotropin protein hormone for use in stimulating secretion of cortisone and other adrenal cortex hormones), hydrocortisone and other body hormones.

Research through the Radon Research Foundation, a non-profit organization, wholly independent of the company operating the Free Enterprise Mine, indicates there are two basic causes of arthritis and other glandular connected ailments:

1. *An external cause:* Stress, strain or shock of physical, mental or emotional origin, followed by—

2. *An internal cause:* Retardation of the gland's activities, resulting in non-normal production of body hormones.

Exposure to the mild radiation from breathing transmuted elements from radon, a gas, coupled with attendant ionization, represents a scientific breakthrough, offering a remedy which reaches a principle *cause* of rheumatoid arthritis and allied glandular afflictions. Reaching the cause, the symptoms of pain and swelling of afflicted joints disappear, as attested by reports of thousands of mine visitors, over a 12-year period of continuous radon mine operation.

Montana's Radon Research Foundation was established in recent years in order to provide a proper vehicle for scientific research. The aims and purposes of this nonprofit Foundation will be hereafter outlined and discussed. No officer or director of this organization has ever received any compensation, or even expenses incident to research. Medical doctors are on its Advisory Board.

Research has proceeded far enough to now confidently announce that radon therapy and ionization appear to represent *one of the most important scientific and medical observations of decades.* Too many thousands of visitors at the Free Enterprise Mine now claim positive results to dismiss their claims as psychosomatic in origin. Likewise, too many individual medical doctors, members of Medical Associations, have been sending patients to the Free Enterprise Mine, or arriving at intervals for their own afflictions. These doctors have checked their patients laboratory-wise, before and after mine visiting, and have noted a precipitous drop in sedimentation rate, often a return of red and white blood cell count toward normalcy, and made other observations and body chemistry tests providing conclusive evidence of physical improvement.

Those medical doctors from many states who have been to Boulder, Montana, and investigated the Free Enterprise radon operation and the uses of radon and its transmuted elements, do not hesitate to recommend nature's treatment to some of their patients. The doctors who have themselves benefited by acquiring rapid freedom from pain and joint swelling, symptoms of rheumatoid arthritis, are thoroughly convinced of the positive benefits that can be attained. Those who were originally skeptics are now the most assured.

The mild radiation, breathed and exhaled in the underground workings of the Free Enterprise Mine, derived from radon and its transmuted elements, has no relation to other types of radiation such as fall-out from the H-Bomb testing, the X-ray, Cobalt 60, and other radiation sources. In breathing radon at the 85-foot level of the Free Enterprise Mine the visitor exhales nearly as much as he breathes in. Therefore, only a fraction of the radon to which he is exposed is utilized by the body. Accordingly, this type of radiation is entirely safe. Mine visiting is limited to a maximum of 32 one-hour visits, two per day, over a 16-day period. Some persons beneficially respond after a fewer number of visits. Those visitors taking less hours exposure usually write in and say, "I have received benefits but feel that I would have benefited much more had I stayed longer."

Initially, the Elkhorn Mining Company, operator of the Free Enterprise Mine, was primarily interested in making certain that radiation underground was well within tolerance limitations; that has been a safe rule to follow, and, based on times of exposure for the casual mine visitors, the company management believes that at all times the company has operated within tolerance limitations. The company is also guided at all times by the technical advice it receives from many reliable sources, both in the United States and from European medical doctors.

Radiation factor units and exposure time periods quoted above refer to the casual visitor only and not to a uranium mine worker who, in working a year underground, might receive 3,200 rather than 32 hours of radon exposure.

A parallel to the Free Enterprise Radon Mine is located near Badgastein, Austria, where the medical doctors of Innsbruck University operate a spa and a 1,500 meter adit containing radon gas, to which their patients are exposed. The time periods of radon exposure recommended at the Free Enterprise Mine are, in part, based on time periods used by the medical doctors at Badgastein. European doctors' written clinical evaluations, records of patient response to radon, have been received from Austria, and scientific conclusions there, with reference to arthritis and other afflictions, are nearly identical to the determinations effected at the Free Enterprise Mine.

Initially, prior to study, and for conservative reasons and maximum safety, fewer visiting hours were suggested to mine visitors. This fact is indicated in the first visitor letters and statements recorded in the chapter entitled "History of the Free Enterprise Uranium-Radon Mine." After an experimental period, advice was received from a Department Head stating that following that Department's trapping the radon by liter container at the 85-foot level of the Free Enterprise Mine and analyzing the amount of radiation in micromicrocuries per liter of mine air, the mine visitor could experience

5

exposure by remaining underground for 30 days, 24 hours per day, without reaching radiation tolerance limitations. Incidentally, there are of course no adverse side effects.

Effectiveness of the radiation utilized at the Free Enterprise Mine is based on the fact that the modicum of radiation acquired by the body results from short half-life elements transmuted from the gas, radon, inhaled and exhaled. Such radiation works from the inside of the body outward, not from the outside inward. External radiation from radium contained in the uranium ores, although present, is not, of course, the type of effective and beneficial radiation employed.

It should be noted that not one uranium mine in hundreds produces a copious amount of radon gas as does the Free Enterprise Mine. Certain mineralogical and geological factors must be present, attending a uranium ore structure, to develop radon gas and maintain it on a continuous daily basis. Those factors are present at the Free Enterprise Mine, which makes it unique in this respect.

The first edition of this book presented more questions than answers. They related to the science of radon therapy and ionization, those factors supplied by nature and beneficial to sufferers of arthritis, bursitis, sinusitis, asthma, skin afflictions and kindred glandular connected ailments. Now, additional research, over the past nine-year period, has resulted in many of the answers. These will be presented in later chapters after first a discussion of the early history of the Free Enterprise Mine.

CHAPTER 2

HISTORY OF
THE FREE ENTERPRISE MINE

When, in 1952, the newspapers and national magazines announced that persons afflicted with arthritis were flocking to the Free Enterprise Uranium Mine at Boulder, Montana, for "cures" from crippling afflictions, many prophesied that here was another "fad" which would die out after a few months. Those claiming almost immediate freedom from pain and years of suffering were classified as psychosomatic cases. One magazine announced that "Barnum was wrong. Two are born every minute." The author was maligned, subjected to vilification, and, through innuendo, accused of fraud. Quoting qualified doctors could have cleared the controversial atmosphere, but such doctors had not given permission for quoting. No one could imagine that "sitting in a mine" for one hour daily could afford relief or "cures" for a long standing disease. No attempt was made at that time by any writer, or source of information, to determine any underlying scientific reason for rapid elimination of pain, decreased swelling of joints, greater mobility of body members. No attempt was made by the author to refute any charges. He was too busy searching for scientific information and reasons which might account for the quick relief from pain, as claimed by so many hundreds of arthritis victims. Twelve years of additional observations and research have determined many reasons.

First came the arthritics, those with painful, swollen joints, some on crutches, others in wheel chairs or on litters. As time passed, hundreds more claimed relief from sinusitis, asthma, and other afflictions. Whatever the cause, beneficial results or cures continued to be claimed by scores daily. Questions arose in the minds of the mine operators. What were the underlying scientific reasons for visitors' claims? Did these claims indicate discovery of a basic cause of arthritis, sinusitis, asthma, and kindred disease? What changes, if any, in body chemistry occurred? Could radiation stimulate the body or the glands to greater activity or normal functioning? Were radiation factors in the mine air safe for human beings and within industrial radiation tolerances? Did the excitement of visiting a uranium mine produce only a psychosomatic response? Did uranium or radon gas, or other elements present in the mine or mine air have anything to do with benefits claimed?

Attempt will be made to answer these and other questions foremost, then and now, in the public mind. Through interviews with thousands of mine visitors, it is clear that people do not need to be trained scientists to ask intelligent questions, and they likewise do not need to be technicians to be interested in the answers. All are interested in an explanation if such be available. Answers to the scores of questions will take scientific study and research on the part of those qualified to answer them; a few will be answered here, insofar as information is available.

The land covered by the present Free Enterprise Mine and installation was originally appropriated under the United States lode location laws in 1924 by miners interested in the silver-lead ore possibilities indicated by vein outcrops at the surface. Prospecting proceeded at a desultory pace over the years and into 1938. Lenses of ore were encountered bearing silver and lead, and a few small shipments of high grade ore were delivered to the local smelter. The miners' prospecting eventually reached a depth of 85 feet at which point the claim was abandoned.

On June 18, 1949, the presence of radioactivity was discovered in the old dump by Geiger Counter, the three locaters being Sanford R. Davis, Edward C. Miles, and Wade V. Lewis. The Elkhorn Mining Company eventually acquired lease, followed by purchase of the property, carried mine development further to three mine levels, and shipped commercial uranium ore to the Vitro Chemical Company at Salt Lake City, Utah, for processing. It is presumed that the uranium thus recovered was eventually sent to one of the Atomic Energy Commission centers for industrial or defense uses. The Free Enterprise group of lode claims represents one of two Montana properties that have produced commercial uranium ores, in which the company is interested.

A few days after location of the Free Enterprise ground and discovery of uranium in 1949, the writer was lowered by rope into the abandoned shaft. A caged canary had previously been lowered to check on the possible presence of dangerous and often fatal carbon monoxide gas frequently found in abandoned mine workings. Canary birds are commonly used by mine organizations and are carried in U.S. Bureau of Mines Railroad Rescue cars for such check by reason of their particular sensitivity to the presence of even a minute amount of carbon monoxide gas.

The writer's experience in his first examination of the abandoned mine shaft was written down at that time as follows:

At the time I was lowered into the shaft, it was soon evident that radioactivity was so great at the station 50 feet down that instrument readings were useless. In five minutes the Geiger Counter, an unsealed instrument, was so saturated with radon gas and its decay elements that the needle exceeded any attempted reading on any scale. I examined the mine workings and the vein structure for an hour and on return to the surface found that my clothing was likewise saturated with radon gas. My hair, hands, every portion of my body and even my breath were extremely radioactive. Quite evidently I had breathed a very heavy concentration of radon gas during the one hour examination. My body did not take the full effect of all the radioactivity to which I was exposed, because I had breathed out nearly as much radon gas as I had breathed in. However, my body and clothing were radioactive for several hours following the shaft trip. (The radon concentration present in the shaft was quite excessive, but has now, of course, been lessened and greatly diluted by mine ventilation.)

Two years after the uranium discovery, and following a period of development and uranium ore mining, a Los Angeles woman, visiting with her husband in the summer of 1951, asked to enter the underground workings of the Free Enterprise Mine. She descended with the author from the surface to the 85-foot level and enroute, while being lowered on a narrow mine cage platform by a steel wire rope hoist cable, admitted that she was afflicted with bursitis. Facetiously the writer suggested that radiation underground might relieve her affliction. Twenty-four hours later her husband telephoned from Helena, Montana, to Boulder stating that his wife was miraculously free from pain with no sign of bursitis. Previously the woman could not lift a kitchen utensil without experiencing extreme pain in her right shoulder.

This Los Angeles woman later invited to the mine a woman friend from the same California city, who was similarly afflicted with bursitis. Three weeks later, after mine visits at Boulder, the second visitor called long distance from California, stating that she had waited those weeks to be certain she was cured, and had even gone to her Los Angeles doctor to inquire the reasons for her complete freedom from pain.

Thereafter the stampede started. Nearly 1,000 people afflicted with arthritis or other afflictions virtually demanded mine visits, and were permitted to visit the underground mine workings. Still skeptical, the mine owners permitted visits for observation purposes only. This work of observation was done, of course, without charge. Percentage response and claims of relief or cures ran so high that the operators, through public demand, were compelled eventually to furnish modern facilities, now installed and available daily to visitors. After nearly thirteen years, following the discovery, visitors have increased, not decreased, and still more continue to arrive at the Free Enterprise Mine from every state in the Union, even from foreign countries.

A history of the Free Enterprise Uranium Mine was reviewed by Wade V. Lewis, President of Elkhorn Mining Company, in a talk entitled "History and Facts Pertaining to the Free Enterprise Uranium Mine," on July 1, 1952, at the Exchange Club, Copper Bowl, Finlen Hotel, in Butte, Montana. Another paper entitled "Importance of Uranium Transmutation Elements," on March 7, 1953, at Helena, Montana, was presented at the joint meeting of the Montana Section of the Institute of Mining and Metallurgical Engineers, and the Last Chance Gulch Mining Association. A condensation of information discussed on these two occasions follows:

The first discovery of commercial uranium ore in Montana, a state historically famous for its mining of other metals, was made in the spring of 1949. Then, a year later, followed the first uranium ore shipments from the Free Enterprise

12

property located near Boulder, Jefferson County, Montana. Discovery of high grade uranium ore near Clancy, Montana, paralleled the Boulder discovery.

Since transmutation elements, known as decay products of uranium, now, in view of recent findings, appear of more importance to mankind than uranium itself, this discussion will dwell more on the importance and uses of these minor transmutation elements.

A score of uranium and thorium ore discoveries have been made in the past five years in Montana, but only a few properties have produced substantial amounts of commercial uranium ores. Ores from both properties, the Free Enterprise at Boulder, and the Wilson-Elkhorn property near Alhambra, Montana, have been shipped for processing to the Vitro Chemical Company, at Salt Lake City, Utah. Several thorium-rare earth prospects have been discovered.

Uranium ores found in Montana occur in nearly vertical veins traversing the monzonite of the Boulder granite batholith. Primary uranium minerals are pitchblende and uraninite. Secondary minerals include autunite, torbernite, troegerite, uranophane and phosphuranylite.

Uranium ores assaying as low as 0.15% U_3O_8 were shipped at a profit to the Salt Lake leaching plant. One percent ore, by the carload, is considered exceptional ore. Smaller shipments of Montana high grade have assayed several percent uranium oxide. Federal premiums allowed, under certain circumstances, through the United States Atomic Energy Commission assisted the uranium miner.

* * * * *

13

At this time, in discussing the "Importance of Uranium Transmutation Elements," I speak only as an individual, and not as an official of the Elkhorn Mining Company. The Free Enterprise represents one of several uranium prospects or properties in which the Elkhorn Mining Company is interested, or has under development or production. At this time I shall for the first time disclose certain findings, and make certain statements for which I will accept full personal responsibility. Those which follow represent observations and conclusions reached over a three year period and relate to effects and benefits of radon and other transmutation elements as claimed by persons afflicted with arthritis, sinusitis, asthma, and other afflictions, after trips into the underground mine workings of the Free Enterprise Mine.

After several months of observing the effects of radon and accompanying uranium decay elements on nearly 1,000 people lowered into the mine at their own request, Elkhorn Mining Company finally completed in June, 1952, installation of a new automatic Otis elevator, designed in San Francisco, housed by a fire-proof building 36x40 feet in dimensions, which includes an office, waiting room and dual rest room facilities. The company at no time had thought of promoting the property as a radon mine. *The afflicted public*, in a manner, discovered it, and *demanded it.* These hundreds of visitors were welcome but finally interfered with the primary uranium ore and mine development to such an extent that it became necessary to either deny entrance to the mine or provide these proper facilities for visiting.

At the mine, each person descends by Otis elevator from the surface to the 85-foot mine level and is assigned one of the underground stations, numbered 1 to 35. Special rooms have been excavated and timbered to accommodate wheel chair and litter cases. The entire 400 feet of drift work, connecting two shafts, is brightly lighted by electricity. Visitors may read or play cards underground. As a safety measure a special auxiliary lighting unit has been provided to function within thirty seconds should regular electrical power fail. Foam cushioned seats and 250-watt heat lamps at shoulder height add to the comfort underground.

A Teletalk system extends from the mine office, waiting room and elevator to all underground rooms and stations, providing constant means of communication between employees and visitors. Total cost of Free Enterprise underground mine development work and present visiting facilities exceeded $100,000.

Every safeguard possible, at this time, has been provided for the safety and convenience of mine visitors. Mine air and gases from all three levels have been trapped by the liter in special devices, and have been analyzed as to micromicrocuries present per liter. Radiation present at the 150-foot level of the Free Enterprise Mine is reported about seven times the intensity of that at the 85-foot level. Mine waters have been checked for possible bacteria, adverse mineral content, and radiation factors, as a safety measure. In visiting any mine the visitor has the right to assurance that these checks and analyses have been made.

Other mine considerations have included

Other mine considerations have included assurance of no carbon monoxide present in the mine workings, and the presence of sufficient oxygen at all times, the latter having required consultation with a ventilation engineer.

Since making these improvements and taking these safety measures, visitors have arrived from nearly every state of the Union and from foreign countries, including England, Greece and Palestine.

At this point I might cite a few instances, among hundreds, of those claiming relief or "cure" from arthritis or other afflictions. Reference to individuals will be by numbers only.

Case No. 365: A St. Paul woman, 65 years of age, an arthritic for 22 years, for 19 years a helpless litter case. Arriving at the Free Enterprise on June 23, 1952, by special rail conveyance, she visited the mine for a series of visits. She had been taking pain pills for the past 22 years. On arrival she could lift her hands only a few centimeters above her prone body; she had not moved her feet in years. After a few days she reported freedom from pain, increased flexibility of body members. In October, three months later, I visited St. Paul, and, in the presence of two of her doctors, observed her sitting up, obviously free from pain, or in need of pain tablets.

Case No. 21: A man 55 years of age, one of the first mine visitors. Prior to his first visits this man could not bear to touch his bare feet to a ribbed rug without pain. Three weeks later he demonstrated that he could dance on concrete without discomfort. On last report he is free from pain after arthritic affliction of 18 years standing.

Case No. 1048: A woman 41 years of age.

Quoting directly: "I visited the Free Enterprise Mine Monday, August 18, 1952. My sinus condition cleared immediately and a serious skin affliction on both hands cleared in four days. A million thanks."

Case No. *2005: A man 50 years of age: "At the time of leaving Tacoma I was afflicted with arthritis, experiencing much pain in body members including ankles, knees, elbows, neck and hip. On the second day at the Free Enterprise Mine I began to get better, pains starting to leave, and now at this time practically all pain has been eliminated from all parts of my body. After leaving the mine and returning to Butte, on the way to Tacoma, I was able to walk without pain, the first relief to that extent I have had in six months. I continue out of pain and can walk any distance without pain or discomfort."*

Case No. *2008: A woman 44 years of age. "Prior to visiting the Free Enterprise I was afflicted with rheumatoid arthritis, experiencing much pain in my right hip, extending to the ankle. I was unable to lie or sleep on my right side during the past six months. I was unable to fully open either hand by reason of the arthritis. After making a series of visits to the Free Enterprise uranium property I am free from pain, able to open both hands, and can now lie and sleep on my right side without discomfort."*

Hundreds of cases could be cited, including those 9 years of age to 90, many claiming relief from pain after 20 to 30 years suffering. One serious case of multiple sclerosis, so diagnosed by the visitor's doctor, claimed miraculous relief after visiting the mine. No claim is made of cure for multiple sclerosis, but many so afflicted have claimed better locomotion and mobility of body members after a series of Free Enterprise Mine visits.

What are the reasons for these "cures" as claimed by hundreds of mine visitors? It might be of interest first to note the approach to the investigation required to answer this question.

Over the past years I have written or interviewed many doctors and scientists high in their professions in search for the answers. All have been co-operative and many have furnished most valuable information. My policy has been never to name a doctor nor quote him directly, unless so authorized. We have, however, many reliable doctors from many states sending their patients for a series of mine visits to the Free Enterprise radon property for relief from arthritis or other ailments.

Initially, I wrote simultaneously several scientists asking this question: "What studies have you made of radon and those transmutation elements following radon in the uranium decay series, and what are the results of such studies?" Some returned fragmental information, others said they had not studied the subject. The assistant to a Nobel Prize winner, the latter in Sweden at the time being presented with a $32,000 Nobel Prize, replied that his radiation laboratory in California had been concerned with effects and changes resulting from intense radiation factors but that no study had been made of the minor short half-life transmutation elements of the uranium decay series. Practically no information existed as a basis for the study of this new subject.

When I visit clinics and doctors I make

clear that I am a geologist-mining engineer and not a doctor. Then I invite attention to the subject of uranium's transmutation elements, natural radioisotopes, described in any modern Nuclear Radiation Physics text. To many persons, the subject appears so new and recent in development that few have touched on this phase of science.

The use of radium, the sixth element of uranium decay, has long been recognized by the medical profession for its external uses. Radium, in turn, breaks down to one other element only, radon, a gas, with an atomic weight of 222.0. The atomic weight of gold, incidentally is less, 197.2. Radon has a short half-life of 3.82 days, decaying rapidly through a series of radioactive elements, five of which, including polonium, lead, bismuth and thallium, have a total half-life of less than 51 minutes. So-called radon "seeds" have their uses in the medical profession. It is interesting to note in the encyclopedia of 1912, that beneficial aspects of radon gas were recognized at that time, over 50 years ago, when the element radon was then known as niton.

Much of the early criticism that developed with respect to claims of "cures" by uranium radon mine visitors stated that all results were merely psychosomatic and, further, that the external radiation from radium contained in the mine ore would be too mild to yield beneficial results to anyone with any affliction. This external radiation is, of course, present, but has never been claimed as beneficial. The high percentage of beneficial results, as claimed now by thousands of cases, has proven, moreover, that such benefits are not due to psychosomatic influences.

A question often asked: "Are there any damaging aspects in breathing a nominal amount of radon for a short period of time, either immediately or after a lapse of time?" This question is in part answered by reference to the case of the original prospector and operator of the mine who sank the original 85-foot shaft, in search for silver and lead, without knowledge of the presence of uranium in the ore and radon in the shaft. He worked there at intervals from 1925 to 1933. This prospector, working without benefit of air circulation, had been exposed to a heavy concentration of radon gas over 27 years ago. At 71 years of age, he was found, through thorough physical examination by a Montana Clinic, to have suffered no ill effects from radon exposure of 27 years ago. This case, of course, relates to a single individual but may supply important information if correlated with many other cases.

Criticism as to dangerous aspects of radiation exposure has not differentiated between radiation results due to radium or radium salts ingested in the human system, and a modicum of radiation encountered through breathing a nominal amount of radon gas, that element immediately following radium in the uranium decay series.

Again the question: "What are the scientific reasons for benefits claimed by hundreds of mine visitors?" The best technical advice received to date discounts effects from external radiation and indicates that radiation, due to radon and other transmutation elements in the mine air of the Free Enterprise Uranium Mine workings, is of such type and amount as to stimulate the master pituitary gland in its production of ACTH, this body product thereupon acting upon the adrenal cortex to produce hydrocortisone, the great pain killer. This reasoning and process apparently account for hundreds of cases afflicted with arthritis claiming first, freedom from pain; second, reduced joint swelling; third, increased flexibility of body members, followed by rapid recovery and normalcy. Recent scientific articles indicate that a score of human ills including arthritis, sinusitis, asthma, skin afflictions and others, are caused by abnormal glandular functioning. In the case of arthritis, relief through doctors may be had through administration of ACTH or cortisone, both manufactured from animal gland products. If, however, the glands of the body may be stimulated to produce the same substances and these body glands may be induced to continue production of these necessary body hormones, then the radiation discovery made in the Free Enterprise Mine workings may ultimately be acknowledged as one of the most important medical radiation discoveries of past decades.

Q. Do the operators of the Free Enterprise claim cures?

A. No, it is unnecessary, because hundreds of the mine visitors do.

Q. Are the operators of the Free Enterprise Uranium Mine practicing medicine?

A. No, utilization of God-given uranium decay transmutation elements, natural radioisotopes, does not constitute the practicing of medicine. Obviously the Hand of the Almighty, the

Great Physician, the Great Scientist, does the work.

Years of study lie ahead. The past few years of pioneering point the way. Having personally observed hundreds of people in pain, then in days or weeks observed them free from pain after years of suffering, I cannot believe otherwise than that this radiation discovery can be more important than the discovery of the atomic bomb, because, while the atomic bomb may be used for the annihilation of millions of people, this new phase of radiation, properly studied, improved, and applied, may be utilized for the benefit of all mankind.

CHAPTER 3

QUESTIONS AND ANSWERS

All new mine visitors ask a number of questions regarding the Free Enterprise Mine. One does not need to be a nuclear radiation physicist to be interested in science. It has been found that every visitor is interested and wants to know the answers. The author does not pretend to be able to supply all the answers but the following are the most common questions and answers given, as based on information available.

Q. Why do visits underground at the Free Enterprise appear to benefit those afflicted with arthritis or other afflictions?

A. The best advice received to date indicates that benefits claimed result from a modicum of radiation acquired in the body by reason of breathing radon gas into the lungs. The radon gas is constantly present in the mine air, being the seventh transmutation or decay element of the parent and original element, uranium. While underground the mine visitor breathes out nearly as much radon as he breathes in. Meanwhile the radon gas which has a half-life of 3.82 days, is also decaying to other very short half-life elements, which, too, are radioactive. These include polonium, thallium, bismuth, and lead. These elements are solids, radon being the only element of the 16 element uranium decay series which lives as a gas. Reliable technical sources suggest that the internally acquired radiation has a catalytic effect stimulating the master pituitary gland to produce ACTH, this acting in turn upon the adrenal cortex to stimulate the production of hydrocortisone, a natural hormone produced by the body if such gland is functioning normally. In other words, the radiation acquired, small in amount, as it is, apparently stimulates the glandular system to activity in the production of body products formerly lacking or being produced in meagre and deficient amounts.

Q. Does the element uranium have anything to do with the benefits claimed?

A. Not directly. Uranium itself before decay is reported not radioactive. The products or elements following uranium in its decay series furnish the radiation which may emit alpha, beta, or gamma radiation. The sixth element in the breakdown series is radium with a half-life of about 1600 years. Radium, in turn, the same element utilized by doctors in certain capacities, is constantly decaying to radon gas. This radon gas thereupon decays to other elements which are reported as effecting beneficial results on arthritis or other afflictions.

Q. How many one-hour visits are required to effect results?

A. Operators of the Free Enterprise Mine learn only by observation. The company keeps records of all visitors and furnishes each visitor, on leaving, a form report designed by a medical doctor, for reporting immediate and long range physical benefits. These reports indicate that usually visitors having only a few one-hour visits do not report benefits as do those who have had 12 to 32 one-hour exposures.

Q. How long do beneficial results last?

A. Many visitors claim permanent relief or cure from arthritis and other ailments, especially those not too long afflicted. It is noted that others claiming freedom from pain, decreased swelling, greater mobility of body members enjoy beneficial results for up to one year, sometimes for years. Others return every year or two, often as insurance against recurrence of symptoms.

Q. Are there any later harmful effects?

A. After nearly thirteen years of operation, no adverse side effects have been observed or reported by reason of Free Enterprise Mine visiting. As described elsewhere, the radon present in the mine workings has been trapped by the liter and analyzed as to radiation factors present to assure safety. Radiation from those elements transmuted from radon in nature is mild and is not to be compared to potent radiation from the x-ray, fall-out, Cobalt 60 or other sources. The Free Enterprise deals not with roentgens but milliroentgens, the latter unit being one-thousandth of a roentgen. Moreover, within the hour underground, while breathing radon contained in the mine atmosphere, *one exhales almost as much radon as one inhales, so* that *only a fraction* of the elements transmuted from radon are utilized by the body. Therein lies the radiation safety factor.

Q. Is there any physical condition for which nature's radon therapy is contraindicated?

A. Yes. Medical advisors suggest that persons with active tuberculosis should not have radon exposure, because radon therapy presumes stimulation of the endocrine glands, and in a T.B. case such stimulation could further activate the disease. The company adheres to this policy.

Q. Are there any special instructions or advice with reference to radon mine visiting?

A. Doctors in Austria suggest at least an hour rest period after each mine visit. Reasons given: The body is adjusting to glandular stimulation. Upon becoming pain free and otherwise benefited, avoid stress and do not overdo in physical activities.

Q. What percent of mine visitors claim benefits?

A. The Montana Free Enterprise management reports that about 80 percent of those afflicted with arthritis claim relief, substantial improvement or cures, where a sufficient number of mine visits have been made. This percentage corresponds closely to the percentage stated in the clinical evaluation reports on the Badgastein Radon operation, conducted by medical doctors of Austria.

Q. What reaction should I expect by reason of a series of sufficient visits in the underground workings of the Free Enterprise Mine?

A. Many people claim immediate relief from

arthritis pains after a few one-hour visits to the Free Enterprise Mine. However, effects vary widely with individuals afflicted with arthritis or other ailments. The Free Enterprise operators learn by observation and through comments of hundreds of visitors. Many in pain report general fleeting pains where previously localized in particular joints, sometimes even a slight intensification of pain for a short time. Thereafter, as reported, the pains lessen, or entirely disappear, swelling recedes, and finally flexibility increases. Improvement has been noted after a few days, or after a period as long as six weeks. Many report improvement from affliction after their return home.

Incidentally, a number of medical doctors who send patients to the Free Enterprise report that laboratory tests show many visitors receive maximum benefits 4 to 6 weeks after mine visiting. The modicum of internal radiation acquired through radon's transmuted elements is presumed to maintain a continuing effect in stimulating the body's glands toward normal production of hormones.

The "fleeting pain" reaction that has been reported by many mine visitors, sometimes occurs within 24 to 48 hours following the first visit. It often awakens one during sleep. No ill effects have been reported from this reaction, but rather a rapid or an eventual physical improvement.

CHAPTER 4

MAGAZINE AND NEWSPAPER
COMMENTS

Soon after making the first observations of
the effect of radiation due to radon and other
radioactive elements on people afflicted with
arthritis, the Free Enterprise Uranium Mine was
besieged by hundreds of sufferers searching for
relief from pain. Dozens visited the mine workings daily and the percent of people claiming
relief or "cures" ran high. Most came from local
Montana Butte and Helena areas. After approximately 1,000 people had been permitted to visit
the radioactive mine workings, and hundreds of
people had spread the word of the new discovery, magazine and newspaper writers descended
in groups to obtain pictures and secure stories
for their publications. The extracts which
follow, taken from newspapers and magazines,
relating to the Free Enterprise Mine, do not
necessarily convey complete accuracy in

observations, but are quoted to indicate the scope of coverage for the public. For the most part reports of this nature have been factual and fair, but it must be remembered that no background of scientific studies was available and much of the reporting had to contain conjecture.

The first extended coverage relating to the Free Enterprise property appeared in a copyrighted front page article in the *Montana Standard* on February 10, 1952. It was written by William J. Clark, reporter for the *Standard*. The kindness and courtesy of the editor and manager of the *Montana Standard* are acknowledged in permitting the following quotations from its news columns:

Could a chance visit to a Montana uranium mine have been the key which eventually may unlock the secret of "cure" for a disease which has stricken seven and a half million Americans — arthritis?

The possibility — and that seems to be all that people in touch with the situation will admit — is raising talk which has spread throughout Montana and reached a number of other western states. Reports of a possible connection between radioactivity and health is bringing a small trickle of hopeful people to this small cattle-raising community of the Boulder valley.

The directors of the uranium mine will say, up to this point, only that the matter is being "investigated." They have advanced no claims, but admit they have received many inquiries from people who say they have arthritis in one of its various forms and that they wish to visit the mine.

* * * * *

Doctors have been asked about it. Some quite frankly are openly hostile; some expressed keen interest; one doctor suggested that if his patients were helped he himself would go down. He has "arthritis", he said. So far he has made no such trip.

The Elkhorn Mining Company has kept mum about the newly sprung medical aspect of its mining venture. Its president, Wade V. Lewis, acknowledges he doesn't know much more about it than what is contained in the statements made by the subjects themselves. The mine is called the Free Enterprise, a half-mile from Boulder.

"We do not, under any circumstances, wish to raise false hopes," he said. "We have consulted leading medical and scientific authorities about it, but so far have nothing to announce officially."

The company, through Lewis, issued a formal statement in February. It read: "Pending continued and thorough investigation of the 'curative' powers claimed by arthritic visitors at the Free Enterprise property at Boulder, Montana, the officers and directors of the Elkhorn Mining Company have decided to permit no further visits at the property for the present, pending construction of improvements and certain mine conditioning necessary to meet all the convenience requirements. The company states, however, that following completion of such improvements and additions to the mine facilities, it will soon thereafter make public announcements.

* * * * *

There were — and are — plenty of skeptics.

The six-foot gray-mustachioed engineering graduate from Oregon State College, a veteran of 12 years of service with the Department of the Interior, added, "This thing could be dynamite. Leave the medicine to the medicos."

It hasn't been that simple. People with arthritis who have heard about this "cure" underground are swarming around. "We can't handle this thing," said Lewis in some distress. "We would like to help, if this be help, but we aren't set up for it. We can't go into the health business. We're a mining company, developing uranium ore under a government contract. We can't go into the health business."

A physicist who had made tests on the mine interior declined to involve himself in the pro and con. The "con" came from people who feared a uranium mine might deliver to them some wholly unrelated condition which might develop later. No doctors have visited the mine; there would seem to be little for them to learn there, anyway. There is an obvious lack of research with the use of animals, or at least an inconclusive amount of information about experiments along this line, if indeed such experiments have been performed.

Lewis was troubled by the stories being circulated. What could he say? That the rumors of "cure" were true? That they were false? That . . . the whole thing might be psychosomatic? He didn't know.

"And we don't know yet," he insists. "We have never made a claim about 'miraculous cures' in the mine for any ailment . . ."

"We do know, however, there is radon gas in the mine. Radon is the element immediately below radium in the field of radiation energy . . . We

may speculate that if arthritics seem to be helped by visits to the mine, their breathing of the radon may have something to do with it. It's mere speculation, and we are investigating."

Lewis is caution personified. At the same time, he appears nettled that there is so little medical and scientific information to go on. The letters he must answer won't permit him to ignore the situation. Everybody with arthritis wants to know.

"Many reports have been circulated concerning the beneficial medical effects resulting from breathing radioactive transmutation products contained in the Free Enterprise Uranium Mine workings," he said in a statement. "These beneficial effects have been claimed by a considerable number of people afflicted with arthritis and now claim relief or cures.

"Neither the Elkhorn Mining Company, operator of the property now being developed under an exploration contract for uranium, nor its officers, make any claim whatsoever, in fact are desirous of avoiding publicity at this time, but it is difficult to control the statements of mine visitors who say they have been in pain for years and who now claim relief and freedom from pain. We now have a considerable list of those making such claims. Our company is interested only in determining what, if any, favorable or adverse effects there may be. We appreciate that results may be partly or wholly psychosomatic in some instances, but we think there appears some reason for favorable effects, warranting further scientific study of the subject.

"Elkhorn Mining Company referred the subject some weeks ago to some of the best doctors, radiologists, clinics, foundations, and radiation

laboratories in the United States, at which sources study is being made, but we are quoting none of these sources. We ask the question: 'What studies have you made of the transmutation elements immediately following radon in the uranium decay series, and what are the results of such studies?' In these radioactive elements, we deal with polonium, lead, bismuth and thallium, these being represented by generic symbols RaA to RaG inclusive. The first five products following radon have a combined half-life of less than 51 minutes. In the Free Enterprise Mine workings these short half-life products are steadily replaced so that in all probability there is being built up a constant in percentages of elements always present.

"It is understood that normal radioactivity of human tissue is due in considerable part to potassium (K^{40}), one of the radio elements. Our company is endeavoring to determine how much radioactivity, however little, is normal for the body, and how much is required for normal functioning of the body. We are asking questions: Do modern living conditions tend to dissipate the meager radioactivity required for normal functioning? Do the mine workings supply only a catalyst effecting favorable results? Do the radioactive break-down elements, though meager in amount, indirectly supply some product essential to body chemistry?

"Uranium is not the only mineral present in the Free Enterprise workings. Other ore minerals present include silver and lead. In fact the first operators mined and shipped ore for such minerals before discovery of its uranium content. Conclusions of most scientists who have expressed an opinion indicate that neither beneficial nor harmful effects should result from the relatively low

34

radiation count and very meager amount of radon present in the Free Enterprise Mine workings. It has been suggested by one prominent scientist, nationally known, that some attendant factor (outside of radiation or mental reactions), peculiar only to the Free Enterprise uranium property, may be involved, resulting in cures claimed by arthritic visitors. This might mean the presence of a catalyst transmutation element effecting a result, or a combination of factors rather than a single factor.

"Studies made so far and reported upon indicate that most attention has been directed to the lethal effects of intense radiation, such as radiation following an atomic bomb burst, while not too much attention has been given to study of transmutation elements following radium and radon in the uranium series. At the Free Enterprise Mine radiation is far below ordinary tolerance count and, under present working conditions, there is only meager concentration of radon gas, the radioactive element following radium. It would appear that dangerous radiation factors should therefore be eliminated.

"The officers of the Elkhorn Mining Company wish to stress at this time that a thorough scientific investigation of the claims is now being made, but that no positive announcement will be made until a preponderance of scientific information becomes available. It is not the desire of this company to raise the hopes of arthritic sufferers that a cure has been found for any types of arthritis, although attention is being given the subject, based on statements of those claiming beneficial results. Later this company will make an announcement following receipt of the best scientific advice obtainable."

Radon gas, if that be it, has no odor or taste.

It has no sensory effect discernible to the person who breathes it, Lewis says. People who have breathed it say the same.

A Missoula woman was disappointed about it. There was no sensation at all, she complained. She had gone into the mine with her knees swollen so she scarcely could hobble, her jaws locked by ankylosis almost so tight that she spoke through nearly clenched teeth. Lewis had a letter she wrote after she reached home.

"On the way to Missoula, just after we were passing through Drummond," the letter says, "my jaws suddenly broke loose. I could work them freely, yank them back and forth. How this happened I do not know. I know my visit though, had everything to do with it, and I praise God I was fortunate enough to come."

Another subject is a Butte police sergeant, who now is 56. He says he has had arthritis since the age of 42. A former street car motorman, and afterwards a bus driver, he hobbled to work the day the Mayor gave him the desk sergeant job. It was a place where he could work sitting down. He visited the mine once — and then two more times.

"Look at my hands," he said. He held up two sets of gnarled fingers.

"Now watch me do this—" and he did a jig and trotted up and down a short flight of steps.

"Do you feel anything?" he was asked.

The sergeant, whose feet had given him nothing but toe aches and heel pains for 14 years, bounded up and down. He pounded his feet on the floor. And then he looked up and said, "Now how do you think I feel?"

There are other stories — testimonials — if you call them that, in a similar vein.

A middle aged building custodian by occupation was taking medical treatment for arthritis in the hips. His acute physical pain was a topic among mining engineers in a Butte building where he worked. One of them told him about the uranium mine, how it reportedly had helped some people.

"I couldn't sleep — sleep was out of the question," he said. "The pain was terrific. It took me a good 10 minutes to get into my car to drive it, and another 10 minutes to get out again. I could go to bed, but I couldn't sleep. Finally a friend told me about the Boulder mine." He shrugged. "What did I have to lose — in pain for months.

"I got into the car one sunny afternoon about 1 o'clock and drove the 35 miles to Boulder. It was about 2 o'clock when I got there and after talking to Wade V. Lewis I was allowed to go down in the mine. I stayed about an hour, I guess. Then I came up, drove home, and my wife and I had dinner about 6:15. When we finished, I got up from the table and walked away.

"My wife said, 'Do you notice anything wrong?'

"I said, 'Did I do something wrong?'

"She said, 'You're walking again.'"

"And it was true, I was walking again. I could always hobble around, after this arthritis got me, but always in such pain. I took treatments for quite a while, but the pain would never leave me for an instant. That's what made it so bad, trying to sleep.

"And then, in a matter of four hours — from 3 to 7 o'clock — the pain disappeared. A miracle? Maybe. Maybe it was a miracle, at that. I know it was the answer to all my prayers.

"When you've had a terrible, terrible pain,

day in and day out, and then you go down in a mine, and then something wonderful happens like this — something you don't know about, can't explain — yes, it could be. If it's not a miracle, what is it?

"I'll never cease to be grateful. My pain is gone."

A Butte printer with arthritis in the knuckles of his hands, the hip, the shoulder and neck, made a 'let's see' trip into the mine.

After one visit his hip and shoulder ailments disappeared. After a second the neck muscles loosened so that he felt no pain there, either. The knuckles still hurt.

"Everything in your joints and muscles, if you've had what I have had for several years, hurts the day you go down in the mine and the next day," the printer explained. The police sergeant says the same thing.

"It makes me awful sick for the next day," said the sergeant, "but I am much better. You should have seen me when I couldn't even wiggle my toes."

The Butte printer stood by. "I hope some day to cure my knuckle condition," he said. "My three visits have cured or helped everything else."

The stories above as taken from the columns of the *Montana Standard* of Butte are no longer surprising. After over twelve years of observation, they still recur from day to day, from week to week. Visitors' claims of relief, benefits, "cures", and freedom from pain parallel each other in similarity. The initial painful reaction for a day or two may not occur, but is commonplace, followed by continued freedom from pain. Reaction to mine radon exposure is usually claimed with the second 24-hour period following initial mine visit.

On Sunday, July 13, 1952, *The Independent Record of* Helena, Montana, ran a full page story on the Free Enterprise Mine discovery and operation, together with surface and underground mine photographs, picturing the Free Enterprise reception room building, its interior, and the 85-foot mine level workings. The story was written by the *Record's* staff correspondent and photographer, Miss Dorothy Helton, who produced a very factual, well written, complete narrative relating to the operation. The author quotes Miss Helton's story, in part, and acknowledges the kindness and courtesy of *The Independent Record* in equitable reporting. The story was entitled "Fabulous Story behind Free Enterprise Mine at Boulder."

What "gives" with the Free Enterprise Mine, operated by the Elkhorn Mining Company at Boulder? It's an interesting story so far and it may be years before the last chapters are written.

Little did the original owner, who dug an 85-foot shaft in search of silver and lead back in 1925, dream that one day the diggings would boast a modern building overlooking the town of Boulder and that hundreds of ailing persons would come there seeking that will-o'-the-wisp, relief or possibly cure for arthritis.

The mine and its pathetic procession of visitors has caused considerable comment, written and verbal. Many have adopted a tongue-in-check attitude about the whole thing; others believe that the Great Scientist is attempting to show man that uranium, which man uses to destroy, may be used to heal; some pass it off as a sort of religious shrine, a mecca for crack-pots whose ailments are all in their head; others imply that the men in charge of the mine are money-grabbers, promoting the boom for their own benefit, and then there are those who have been relieved of pain or whose friends or relatives have been benefited.

A person cannot visit the mine and talk with those in charge without becoming aware of their sincerity of purpose. Officials, primarily Wade V. Lewis, president of the Elkhorn Mining Company, have gone all-out to get to the bottom of the mysterious boom. They plan to carefully check between 5,000 and 10,000 persons before issuing any statements regarding the curative properties of the mine.

At no time has the Elkhorn Mining Company claimed any aspects of the Free Enterprise Uranium Mine as beneficial for any affliction. There's no stopping the excitement of visitors who find relief, however, and for nearly two years the mine has been opened to scores of arthritics who heard of it by word-of-mouth.

"Until we know more, we can't issue any statements," Lewis said. "We're hardrock miners, not doctors. We have invited the criticism of doctors and urge them to carefully examine their patients both before and after visits to the Free Enterprise. We're keeping permanent files on each visitor. Our nurse notes signs, symptoms and improvements each day. All information thus acquired is confidential and becomes part of the visitor's permanent file. We can supply physicians with complete information now or at some time in the future, because we plan to follow each visitor with an inquiry three months after the visit and they will be asked about their physical conditions for a one to two year period after leaving the mine."

He explained that the Elkhorn Mining Company had invested more than $100,000 in mining development and new equipment because, since the word spread, people were arriving in such numbers that they interfered with the primary purpose of the mine, the extraction of uranium ore.

Lewis has adopted a "wait and see" attitude and is meanwhile conducting research and compiling case histories of all visitors.

The mine first struck the headlines when the first car of commercial uranium ore was shipped from it in March, 1950. The presence of uranium minerals was discovered through use of a Geiger Counter.

In June, 1951, however, a Los Angeles mining man and his wife visited the mine. That chance visit was to change things almost overnight, not only for the Free Enterprise Mine, but for its owners.

The man's wife, reportedly suffering from arthritis, had found that by the following morning she had complete freedom from pain in her shoulder and could raise her arms without difficulty. When she returned to Los Angeles, she told a friend about the amazing "cure", and the second woman visited the mine with similar results. The pain of her arthritis left her and she had greater flexibility of body members.

The word spread over Montana, then interstate and finally visitors were arriving at the Free Enterprise by the scores.

The directors decided that they had two alternatives: close the mine to visitors, or install proper facilities for them. Those who visited the mine *were almost pathetically grateful for the relief they found and it seemed a shame to deny them the privilege of coming there.*

A new, specially-designed Otis elevator was installed, housed in a modern cinder-block waiting room, with picture windows overlooking the Boulder valley and the town of Boulder. Two rest rooms, an office and the waiting room are included in the 36 x 40 foot building.

At the present time it appears that relief and benefits claimed by mine visitors is not due to external radiation, but rather due to nominal radiation contained in the mine air and gases, including radon or those transmutation elements which follow radon in the uranium decay series.

"Beneficial results claimed may, however, be due to a third factor, X, as yet undetermined, and which may be peculiar to the Free Enterprise Uranium Mine," President Lewis says.

Research and studies are being continued with reference to all factors present. Factor X may, as suggested by one scientist, attend the radiation but may be independent of it. Whatever the cause, a very high percentage of those claiming affliction through arthritis have claimed freedom from pain and relief following a few visits to the mine. One explanation of one scientist indicates that the radiation present in the Free Enterprise property may be present in such type and amount as to be sufficient to release through stimulated glandular activities, certain body products benefiting the course of arthritis.

Wade V. Lewis, president of Elkhorn Mining Company, has listed several examples of case histories, representing visitors to the Free Enterprise Mine.

A 54 year old rancher, had suffered crippling arthritis for 30 years, since he was 24 years old. He was unable to step from the street to the sidewalk curb and had expended $50,000 on his affliction. After two one-hour periods in the mine in January this year, he had fleeting pains for 10 days, then complete freedom from pain. He gained 17 pounds in 60 days and regained flexibility of afflicted members and is now able to perform any manual work required of a man of his age. He was checked by a Montana clinic doctor before and after mine visits.

Physical examinations of more than 1,100 uranium mine and mill workers in Colorado, Utah, New Mexico and Arizona showed no evidence of health damage from radioactivity, according to a recent report by a public health service. The studies have been going on since 1950 and will continue for several years.

Officials of the Elkhorn Mining Company have taken every precaution to insure the safety of visitors.

Mine waters have been checked as to bacteria and mineral content, and for radiation factors which might be present. Visitors were not permitted to drink the water until final reports had been received.

Other mine assurances have included assurance of no carbon monoxide present in mine workings.

Mine air and gases from all three levels (85, 105 and 150 feet) have been trapped in special devices and analyzed. Radiation at the 150 foot level is about seven times the intensity of that at the 85 foot level.

On January 29, 1953, about eight months

after the *Helena Independent Record* story appeared, the *Interior News* of Smithers, B. C., Canada, featured an article headed, "Local Residents Get Relief From Arthritis," simply and factually told, reporting statements of Canadian people who claimed immediate benefits after visiting Montana's Free Enterprise Mine. A part of the story with individual names deleted follows:

An almost miraculous relief from the pain of arthritis has been enjoyed by two local residents in the past month or so, following visits to the Free Enterprise Uranium Mine at Boulder, Montana. Both have returned entirely relieved of the pain from which they had suffered. Mrs. L. S., who was afflicted in the shoulder, arms and legs, was the first to make the trip, followed in recent weeks by Mr. R. W. C., a sufferer from arthritis for the past seven years.

Mrs. S. went to Boulder the latter part of December and visited the mine six times. In the short space of three weeks she returned home walking normally, used her arms freely, and no longer suffered pain in her arms and shoulders.

The change in Mrs. S., and a report of her "miraculous" cure which appeared in the News, aroused the interest of Mr. R. C. well-known local resident and business man. The use of drugs had provided him with the only relief from the continuous pain of the painful and crippling disease. Accompanied by his wife he visited the Uranium Mine in the hopes of a cure. Shortly after his first visit to the 85-foot level, where arthritic sufferers are accommodated, a definite reaction set in, followed by the cessation of pain after further visits. He made eight visits to the mine, and since then the pain has disappeared, the drugs have been discontinued, and he can use both his arms freely. Immediately upon his return home he was back at his desk at his office.

These results may give fresh hope to other arthritis sufferers in the district. Both parties have been personally interviewed by a representative of this newspaper, and no attempt is made to sensationalize the facts as given.

The address of the mine is: Elkhorn Mining Co., Free Enterprise Uranium Mine, Boulder, Montana. Living accommodations are excellent and not unreasonable.

(Author's note: On February 6, 1953 Mr. R. W. C. wrote, "I'm improving every day, no return of pain.")

Nearly a year later, following the *Helena Independent Record* report, the *Great Falls Tribune* in its Montana Parade Section on Sunday morning, April 26, 1953, devoted seven full pages in narrative and pictorial report on "health" mines in the Montana area. Following is a portion of the narrative by the *Tribune* Staff Writer:

The West's strangest mining boom continues to roll merrily along on the eve of its second anniversary. This is the rush of arthritis sufferers seeking relief by spending a few hours in one of the radioactive mines that abound in Jefferson County.

While clinical evidence still is lacking as to benefits to be obtained from such mine visits, many who were skeptical a year ago now are convinced that "these mines must have something."

Meanwhile, health-seekers continue to flock to Boulder. Some travel by airliner to Helena or Butte, then to Boulder by bus. Others drive their own cars, many with house trailers attached. Two brothers from Wyoming arrived recently in a housecar made from a converted school bus.

The "health-mine" boom began in the spring of 1951 when a Los Angeles miner and his wife visited the Free Enterprise Mine near here. The Free Enterprise then was producing commercial uranium ore. The wife of the California miner suffered from arthritis. The morning following her visit to the mine, she discovered that the pain in her shoulder had disappeared.

Upon her return to Los Angeles, this woman (whose name has not been made public by owners of the Free Enterprise) told a friend who also suffered from arthritis of her "miracle cure." This friend, according to a brochure published by the Elkhorn Mining Company, "made a special trip to the Free Enterprise." Shortly afterward she reported the "elimination of all arthritic pain and the return of complete body flexibility." The health mine boom was under way.

Meanwhile, other health mines having sprung up in Jefferson and Lewis and Clark counties; or rather, long in-operative mines have been converted into health mines by installation of electric lights and benches. The Free Enterprise remains the most elaborate health mine in the area and is the only one in which health-seekers are lowered into the mine by an electric elevator. The others are tunnels, into which invalids may walk, if they are able, or be wheeled deep into the tunnels in wheel chairs.

No sweeping claims are made by any of the health mine operators. "The Free Enterprise — Is It A Gateway To Health?" asks the Elkhorn Mining Company's brochure. "Is it possible that uranium, which mankind uses to destroy may also be used as a means of healing?"

"People are skeptical when they first come to the mine," says another paragraph in the brochure, "but as soon as they see the results others have obtained and get relief themselves the skepticism vanishes and they spread the news to their friends. The fact that after a few visits to the Free Enterprise Mine, persons show great improvement and freedom from crippling pain astounds most people. Hundreds of testimonial letters speak words of gratitude for the help received."

One theory for the relief given arthritics after visits to these mines is that radon gas, released by disintegration of uranium ores, somehow acts as a catalyst in dissolving calcium deposits in the joints.

Indications that some of the mine visitors obtain at least temporary relief are seen in the fact that many return, from points as distant as Los Angeles, for a second series of trips into the mines.

Even if radioactive properties of these mines cannot cure arthritis, the pure mountain air and baths in the hot mineral springs at Boulder and Alhambra promise restful nights and temporary alleviation of pain. Thousands of arthritis sufferers, it appears, consider even that much a bargain worth traveling hundreds of miles to obtain.

One national weekly magazine was not so kind in reporting the Free Enterprise Mine. The following is caption and narrative, in part, from the issue of *Time* magazine, dated July 7, 1952:

MIND, BODY AND MINES

Suffering man's infinite capacity for self-deception was demonstrated last week, with radioactive trimmings appropriate to the atomic age, in the little Montana mining towns of Boulder (pop. 1,017) and Basin (population 250). From far and near came hundreds of bent, gnarled and crippled men and women, mostly victims of some variety of arthritis, all pathetically seeking a magical cure. Many thought they were benefited. Undoubtedly benefited were the owners of two abandoned silver mines, hotel and motel keepers, beanery proprietors and taxi drivers. Boulder and Basin had not seen the like since the bonanza days of the 1890s.

The rush began last summer, after a visiting mining engineer took his wife down to look over the Free Enterprise Mine, near Boulder. She had such a severe case of bursitis that she could not lift her arm. But two days after the half-hour trip down the mine, she felt better and proclaimed herself "cured." Her husband figured that radiation from uranium ores was responsible. Soon they were back with a friend who suffered from arthritis. After an hour at the mine's 85-foot level, she too felt better.

Invalids Assay High. The word spread through the mining country and so many visitors arrived at the Free Enterprise that they got in the way of the miners who were trying to find out whether its uranium veins were worth working. Wade V. Lewis, an experienced hard-rock miner and president of the company that owns the Free Enterprise, soon discovered that the visiting invalids assayed higher than anything in the mine: every carload was a payload.

Lewis asked medical authorities to check whether there was anything in the mine that could do anybody any harm — or any good. Assured that a trip down the mine should not actually hurt *anybody, but without waiting for a verdict on the reported cures — no serious investigation has begun yet — the owners of the Free Enterprise floated $100,000 worth of stock and closed the mine for improvements.*

A new shaft was sunk, and into it was built an Otis elevator big enough to hold stretcher and wheel chair cases. This cost $50,000. Airlocks were installed in the mine to seal in "curative gases." To keep the procession of health-seekers in order, there is a flossy reception room where each visitor gets a number assigning him to a seat in the 85-foot lateral.

. . . Proprietor Lewis, flanked by a lawyer is careful never to use such words as "treatment" or "patients" . . . says Lewis, "We're mining men, developing a uranium-bearing deposit. We're not doctors and don't pretend to be." But even with a daily limit of 30 new visitors, the mine takes in as much as $3,000 a day and nobody has seen any trucks of uranium ore coming out.

From the start, the Free Enterprise got the carriage trade. The overflow and those who cannot afford its rates go to the Merry Widow in Basin, nine miles away, where the crowds are even bigger. The Merry Widow owners need no elevator because theirs is a lateral shaft. They lead their visitors a steep climb up a rocky trail to get to the tunnel entrance, and levy no fixed charge . . .

There is nothing in medical or nuclear science to suggest that the tiny, harmless doses of radiation, absorbed by visitors can effect joint diseases. Nevertheless, a lot of people who go down the mines feel better afterward. It is also true that some people with pains in their joints (from either disease or a state of mind) feel better after carrying potatoes in their pockets. There is no reason to believe that sitting in a mine is any less effective for such cases than carrying a potato. But in Boulder and Basin it is more expensive.

A bit rough, shall we say. Now for the truth. First of all, the operating company did not invite arthritics to the mine. The first few months they came by the hundreds voluntarily and between 800 to 1,000 people, many in extreme pain, visited the mine. It cost money and time to handle these people but there was no charge. Suffering humanity besieged the writer's home night and day asking permission to enter the mine workings and all were accommodated.

It is true that a total of $100,000 was expended at the Free Enterprise in mining development, in purchase of an Otis elevator, its installation, and attendant facilities. Montana state safety laws required complete safeguards. Incidentally, the author has to date never received one cent in dividends from the Free Enterprise operation, either from uranium ores shipped or from visiting fees later charged. Moreover, he has never, as president and later as vice president of the company, received a salary as such. He was paid nominally for geological engineering services.

The above magazine writer used the term "lateral shaft." As a geologist and mining engineer operating since 1913 the writer has heard of an incline shaft or vertical shaft, but never of a "lateral shaft." Horizontal mine workings are usually referred to as tunnels, adits, or drifts.

As to uranium production, the Free Enterprise Mine produced the first commercial shipments in 1950, the first ever mined from the

State of Montana.

When a Nobel prize winner the same year advised that, although his work was concerned with certain phases of radiation in a California laboratory, he was not qualified to comment on reactions reported at the Free Enterprise Mine, by reason of the presence of radon gas and other natural radioisotopes in the air of the mine workings, it certainly ill behooves the magazine reporter to comment adversely on a subject requiring scientific study and research. The comments, if properly presented, would have required the experience of experts in medicine and radiophysics.

Life magazine of the same date, July 7, 1952, devoted three full pages to pictorial and narrative reports regarding the Free Enterprise Mine and a neighboring property. This reporting was in a kindlier tone and endeavored to report the facts. It read in part as follows:

Near the little Montana mining town of Boulder last week a couple of second-rate uranium mines were busily marketing a new and profitable stock in trade. Their commodity: hope. Their customers: sick and suffering people desperately seeking an end to their torments. Hobbling on crutches or borne on stretchers, victims of arthritis and other chronic diseases came from all over the West, lured by word-of-mouth reports of miraculous relief to be found in the radioactive air of the deep shafts.

The first of these reports came from the wife of a mining engineer who visited Boulder's Free Enterprise Mine. Upon emerging she discovered that her left arm, long immobilized by bursitis, could move again without pain. "If you have suffered for years with tears in your eyes," she told her friends, *"you'd realize that this was a wonderful thing." The word spread. By March of this year 750 people had come to the mine and gone away claiming improvements. Said one, a bus driver whose arthritic hands had become usable again: "I don't know how I'd have made a living much longer if it hadn't been for the mine." It was then that the mine owners decided to suspend mining operations and install a waiting room and elevator for their pathetic pilgrims . . .*

While the Free Enterprise was being converted, its owners sought an explanation of the mine's alleged curative powers. The experts informed them that the radioactive radon gas which the mine contained could not help anyone. In view of this the management avoided claims of any sort . . .

Among medical officials, who view the Montana "miracle" with scientific detachment, the consensus of gloomy opinion is that "Barnum was right"…. But so great is the power of hope that only a handful of the 5,000 who have passed through the mines deny that their visit has done them any good.

Two of the pictures accompanying the above magazine narrative showed Mrs. E. K. first in a wheel chair, second walking from the Free Enterprise waiting room. Captions were as follows:

Waiting, E—K—, chair-ridden, breathes gas in mine. She predicted she would walk again.

Walking, E—K—, leaves the mine after third visit. Before visit she said she could barely stand up.

In April of 1954, a correspondent of the *Saturday Evening Post* describes his personal battle with rheumatoid spondylitis, or rheumatoid arthritis of the spine, and details the modern methods of treatment he received over a long time period for obtaining relief to the extent that he can walk and engage in his usual daily activities. The magazine quotes him as convinced he could not be cured, but was happy simply to have the disease stopped in its course, or at least, in treatment of symptoms, be given all the therapeutic relief and benefits available through modern medical science. His article was one of deep appreciation for help as one of "10,000,000" arthritis victims in the United States. The prescription noted for this particular case included aspirin, diet factors, physical therapy, X-ray treatments, and, in emergency, pain tablets, all administered under the direction of the best doctors.

The article noted that unfortunately less than 1% of those afflicted with arthritis receive any kind of hospitalization, in most cases few arthritics can afford hospitalization the disease eventually impoverishing most victims.

Prior to receiving valuable modern treatment friends recommended many unorthodox remedies including bee stings, leeches, copper bracelets, radioactive rocks in one's bed, or visits into the underground workings of old uranium mines. The last statement presumably refers indirectly to the original discovery of the Free Enterprise Mine at Boulder, Montana, and unwittingly includes that mine in the same category as the dozens of hoaxes of "radioactive uranium mines" and "inventions," promoted by unscrupulous persons aping the original discovery of the Free Enterprise Mine. Elsewhere in this story this unfortunate situation in interstate activities is detailed.

Sufficient to say at this point that "uranium" frauds or misguided activities should not be included in the same category as a radon producing mine property, that there is a vast difference between the element uranium, a solid, and radon, a gas, and that warranted criticism from any source would require combined experience of radio physicists and doctors with a radiological background.

No claim is suggested that a series of visits to the 85-foot level of the Free Enterprise Mine would grant relief or benefits in the particular case of the last mentioned afflicted magazine correspondent, but stranger things have happened, and an invitation to visit the Free Enterprise Radon Mine, based only on his doctor's permission (not approval), is now extended, with, meanwhile, a good thought being held for his continued physical improvement.

More recent criticism, direct and indirect, of the Free Enterprise radon operation has continued over the years. Such adverse criticism has come from Foundations and National magazines that should know better. Their staff writers wrote without investigation or research. The *Reader's Digest*, of June, 1960, contained an article by Blake Clark, entitled "The Cruelest Swindle in Medical Cures."

By context inference, page 157, it included the authentic Free Enterprise Radon operation in discussion of obvious frauds. Upon publication of that issue the *Reader's Digest* received so many letters from persons who had visited the Free Enterprise and received benefits, that the *Digest* management finally sent out a form letter in reply, rather than individual letters. No retraction of statements has, to author's knowledge, been made. Meanwhile, many persons, who might have become pain free, accepted the Blake Clark article as authentic and continued their suffering. Some determined the truth later. Other national magazines have likewise been misguided in publishing articles wholly inaccurate, relative to the Free Enterprise operation. Attention is directed to a book by one Ruth Walrad, "Research Consultant," copyrighted 1960, by the Arthritis and Rheumatism Foundation, New York, and entitled, *The Misrepresentation of Arthritis Drugs and Devices in the United States. A* Foreword, dated May, 1960, is written by Ronald W. Lamont-Havers, M. D., Medical Director of the Foundation named.

While the foregoing book does contain factual information relating to fraudulent devices, it has made a very serious mistake in inferring that the Free Enterprise Radon Mine is fraudulent. I quote from pages 78 and 79 of that book:

A very different type of treatment is that offered by uranium tunnel operators who have so grossly exploited the vulnerability of the arthritic to false claims. The original mine, located in the West, received wide publicity from a two-page illustrated article in the July 7, 1952, issue of Life *magazine. An unsuccessful uranium prospector devised the scheme of turning his losing venture into a bonanza by promoting the "radiation" from uranium dirt as beneficial to arthritis. For $2.00 an hour one could sit in the mine tunnel, supposedly receiving emanations from the ore.* Life *undoubtedly meant to expose a sensational swindle but it opened a veritable Pandora's box. Other* *operators of little conscience adopted the scheme, and arthritics, wanting to believe in the magic cure, made it profitable. The amount of radiation in the mines has been checked and found to be undetectable, negligible, or at the most, about that received from an illuminated watch dial. This is the one fortunate aspect of the* whole swindle, *for otherwise patients might have been physically harmed.*

In areas unendowed with uranium, low-grade ore has been shipped in to line walls of abandoned stores, old mine tunnels, or even to put in troughs where crippled feet could be "bathed" in the so-called healing dirt. Most of these were crude structures but a few promoters gave added conveniences and comforts. One in Montana invested $75,000 in an elevator and other improvements, and some have provided music, reading material, and other means of whiling away the time while absorbing the radiation.

Government agents, both federal and state, have taken quick action against the operators. California, for example, has eliminated six in the state and has a seventh before the courts. For a time the craze subsided but it appears to be returning to popularity.

The author, a graduate mining engineer, is apparently referred to as "an unsuccessful uranium prospector." The statement "one (mine) in Montana invested $75,000 in an elevator and other improvements" evidently refers to the Free Enterprise Radon Mine, the only known radon operation in the United States with an Otis elevator. In the opinion of the author this book should either contain an addendum retraction or should be suppressed from the United States mails for misrepresentation and false statements, so far as the authentic Free Enterprise Radon Mine is concerned.

Recently, through statements originating in California, attributed to a San Jose assemblyman, and reported by radio, television and news dispatches, there has been questioning as to the authenticity of claims that underground visits to the Free Enterprise Mine at Boulder, Montana, can be beneficial to those afflicted with arthritis and allied glandular connected ailments.

This assemblyman asserted, according to news media, that "a pamphlet put out by Elkhorn Mining Company may violate California laws," and suggested that the state's attorney investigate the Montana "uranium cure mine."

Such adverse criticism is wholly unjustified and is based on abysmal ignorance of radiation physics and ionization. The Free Enterprise Radon Mine does not deal directly with radiation emanating from uranium ores and has never claimed that uranium per se benefits arthritis or other afflictions. Rather, nature's therapy is based on the breathing and exhaling of radon, a gas, an element derived directly from radium contained in the ores. Radon, in turn, decays to 9 other elements, the first 5 having a half-life of less than 51 minutes. Radiation utilized by the body works from the inside outward, not from the outside inward. Scientists conclude that this radiation, mild and safe for the few hours of exposure, does, by direct or catalytic action, stimulate the boss pituitary gland in its production of ACTH.

Doctors assert that ACTH acts upon the adrenal cortex, stimulating the production of hydrocortisone, the body pain killer.

In brief, radon therapy, as provided by nature, with its important attending factor of ionization, tends to bring the body glands toward normalcy in production of body hormones.

As to "investigation" of the Elkhorn Mining Company Free Enterprise Radon operation. It was investigated over 10 years ago by several Federal Government agencies. We quote none but do say that all such agencies have been most gracious in considering and studying the scientific factors relating to the Free Enterprise Radon operation.

The Free Enterprise invites and welcomes full investigation of its activities, by State and Federal agencies at all times. Our scientific research is continuous.

Sciences concerned with the Free Enterprise Radon Mine, now in its 12th year of continuous operation, include mineralogy, geology, radiation physics, ionization, uses of radon's transmuted elements, and their effects on the endocrine gland system.

The first edition of this book included the chapter entitled: "Frauds You Will Encounter." It pointed out that not one uranium mine in hundreds produces a copious amount of radon gas, the effective element.

To clear the record: Many individual medical doctors now send some of their patients to the Free Enterprise Radon Mine, or arrive at intervals for their own afflictions. They agree that radon therapy causes no adverse side effects as in the administration of cortisone.

The Elkhorn Mining Company always invites criticism of its Free Enterprise Radon operation, but it also invites study and research by those qualified to conduct it.

Montana now has a Radon Research Foundation, a nonprofit organization, with Medical Doctors on its Board of Medical Advisors.

With reference to California or other criticism of the Free Enterprise Radon Mine, may we quote Barney Baruch who once said: "Anyone has a *right* to an opinion, no one has a *right* to be *wrong* in relation to the facts."

CHAPTER 5

WHAT THE MINE VISITORS SAY

It is well understood that the medical frater-
nity by reason of its careful, scientific approach
to new matters, must of necessity accept only
findings that are substantiated by detailed clinical evaluations. However, statements voluntarily
made by many arthritics and those afflicted with
other ailments contain an element of interest
and certainly reflect the sincerity of the public,
even though they represent self reporting and
self diagnosis. The letters and statements which
follow are quoted in whole or in part. They are
assigned numbers for identification, which numbers are not necessarily mine visitors numbers,
but which may, at any time, identify the particular individual for record purposes. Initials
rather than names are given, with street addresses eliminated. These testimonials were received during the first years of radon mine

operation. Statements are deleted which might subject the mine visitor to undesired publicity or subject others to criticism.

Since these early testimonials from Free Enterprise Radon Mine visitors were volunteered, thousands of other afflicted people have passed through the mine portals. Statements of visitors of later years are added. Many of these reported benefits on a form prepared by H. R. Tregilgas, M. D., of South St. Paul, Minnesota.

One of the earlier cases of Free Enterprise Mine visiting is particularly interesting in view of a first report ten days after the initial visit and a second report about ten months later. The case concerned a Butte, Montana man.

Case No. 5: Upon my arrival home after my first visit in the Free Enterprise Mine for one hour, Thursday, November 1, 1951, my wife, after our dinner, noticed a great difference in my walk. The next day there was more improvement, I walked with ease, and faster. Friday night, the first time in over three years, I was able to slip off my casual shoes and remove my clothing.

After the next trip down the mine for one hour November 7th, I felt still better. I was able to start walking at once upon arising from bed or a chair. Before, I had to stand for some time before I was able to move. I can bend and touch below my knees with my fingers. I just can't explain the marvel of it, I feel wonderful, can't really explain just how I do feel. It's so wonderful, to be relieved of such long suffering, the thought of being crippled for life, and unable to work. God knows — the very thought of being able to enjoy good health, and live again, is an answer to our prayers, and thanks from the bottom of our hearts,

miracle.

Note: On her first day of arrival at the Free Enterprise Mine, November 23, 1951, this young woman walked with a cane; she reported that she was consuming large quantities of drugs, by reason of extreme pain, that her jaws were partially locked by reason of ankylosis. On last report, over two years later, she was active as a beauty parlor operator.

It is interesting to note the report of this young Montana woman, referred to as Case 58, nearly a year later:

For the past six years I suffered from arthritis. I have gone to many places, such as Wheeler, Oregon, and the Mayo Clinic at Rochester, Minnesota, and have done many things to try to relieve it. Everywhere I went I was told that it was hopeless and that I wouldn't get any better, I took high amounts of aspirin — up to 100 tablets a day — and any other pain-killer I heard of. I don't know how to describe the pain I had all over; I only know it was terrible.

I finally had to use a cane to get around. My arms were so sore and I was so weak, I could hardly manage the cane. In the mornings it was agony to put my feet on the floor and I could only shuffle them along. I also had a bad sinus condition that was very aggravating.

By the fall of 1951, I was so much worse I had about given up hope when I heard about the uranium mine at Boulder, Montana. As we had tried everything else, we decided to go there, making my first visit to the mine the day after Thanksgiving, 1951. I was so bad that day I went in the mine I could hardly stand, and if my husband had not helped me, I doubt if I could have made it that morning.

About four hours after we came out of the mine, when I started to talk to my husband, I felt a cracking or breaking in my jaws. By the time we arrived in Missoula my legs were cracking. I was very tired and went to bed and slept all night.

After that, I began to get sore and more movement in my joints. In two weeks I went back and made two more trips in the mine, and then waited a month and made another trip.

After the first trip down, I noticed an improvement in a sinus condition I had, that I had since the arthritis started. I have been steadily improving since that first trip. I became stronger and more active, and was able to do more. Finally, I did not need the cane any more. I hadn't been able to drive the car for years, and since I could handle my legs and arms, I have been able to do that also.

I am a beautician and I began doing some of the work, finding that I could stand longer each time and am now well enough to be working in a beauty shop again, standing on my feet long hours and working hard.

I know I do not look or feel like the same person that went down in the mine last November, and I seem to be getting better all the time. After the trips in the mine the pain gradually left. There were days when it would seem worse and then it would change and finally, the last place I had pain was my knee. When that pain left, I really began to get better.

Case No. 132: That affiant is 54 years of age, has been afflicted with arthritis since the year 1922, for a period of approximately 30 years, that present doctor's name is ——, of ————, said doctor having been familiar with my case for the past 12 years.

That on January 8th, 1952, affiant was still afflicted with arthritis, with all the attending factors of soreness, lameness, stiffness of joints, including pain in shoulders, elbows, knees, neck and back.

That affiant made two trips for one hour each to the 85-foot level of the Free Enterprise Uranium Mine near Boulder, Jefferson County, Montana, on January 8th, and on January 9th, 1952.

On return home to Three Forks, Montana, affiant noted painful reaction for about 10 days, then slowly began to get better, and then gradually all pain was eliminated, body flexibility increased and was thereafter and is now able to do heavy lifting which he had been unable to do for the past 15 years.

Prior to visits to the Free Enterprise affiant was limited as to diet, but now eats normally with a definite increase in appetite; that affiant, after affliction since 1922, is now rapidly becoming normal. Circulation has vastly improved.

In April, 1954, 27 months later, affiant, a Montana rancher, reported to the Boulder office that he still claims freedom from pain, shows continued improvement, and that he is active daily in ranch duties.

Case No. 682: My sinus condition cleared immediately. A serious skin condition on both hands cleared in four days. A million thanks.

The Free Enterprise Mine employees saw the condition of my hands the day I visited the mine, so they know how badly I felt. It is with sincere thanks that I can say I benefited greatly. I told

Mr. T. this same thing, as three days after my visit, my hands were clear of any eruptions. I had consulted a skin specialist a few years ago when I had an attack; also followed orders, carefully, given each summer when I get these attacks, so was hesitant to do something as easy as sitting in a mine for one hour. I was there on Monday noon, August 18th, and by Thursday, August 21st, my hands were completely clear. I had one siege since, but it was not convenient to get to the mine again.

I have told many of my friends of the help I received in that one visit. July and August is the time I have trouble with eczema so I hope the sinus stays clear too.

Thank you again for the help I received.

Mrs. A B., Montana — November 25, 1952.

Mrs. M. W. in voluntary deposition, subscribed to March 27, 1952, at Butte, Montana, reported as follows:

Case No. 62: I wish to express my gratitude to you and your men for allowing me to go down at the Free Enterprise Mine. It has done wonders for me.

I first felt pains and aches about ten years ago (1942). These pains and aches existed in my toes, ankles, knees, hips, spine, neck, shoulders, elbows, wrists and fingers. I also experienced terrific pain in my jaws, causing a locking and snapping.

I doctored with a doctor at ——, Montana, for at least six to eight months with no relief. I then went to ———, Wyoming, for about three weeks. No relief was felt from these treatments either. These treatments consisted of hot baths, massage, and rest.

I hadn't attended Mass for three years. I was

not able to kneel down. Stiffness in hips, knees and ankles, unable to climb stairs, or even walk without assistance. I was unable to grasp a broom or any object any smaller. I was never able to sign my name to any papers as I couldn't hold a pencil or pen. Housework was out of the question. I could not stand the weight of the bed covers on me at night, no matter how light they were. My daughter had to lift my head and place my pillow under it. Many times I was unable to get in or out of bed without assistance. Riding in a car was impossible.

On February 28, 1952, I made my first trip to the Free Enterprise Mine, in Boulder, Montana. I went down the mine at 2:00 P.M. on said date, and I sat down there for one hour on the eighty-five foot level. The next day, February 29th, at 10 A.M. I called my daughter-in-law to tell her I had no more pains or aches. I was able to do everything that I was unable to do previous to that time.

My pastor and neighbors will verify the change they saw in my condition since going down in the Free Enterprise Mine.

I will be happy and glad to answer any questions concerning my health as it was before, and as it is now after going over to your mine. My deepest gratitude and prayers will always be with you.

Case No. 704: That affiant is 62 years of age, has been afflicted with rheumatoid arthritis since February 1952; his doctor being Dr. E. whose address is Medical Dental Building, Everett, Washington; that on the 17th day of February, 1952, affiant was afflicted with clot in the arteries, causing partial paralysis of the left side; that shortly thereafter he was afflicted with rheumatoid arthritis, with attendant swelling of hands, knees and ankles, considerable pain, together with other symptoms attending the arthritis.

That the foregoing represented the physical condition of affiant on and after Wednesday, October 22, 1952, at which time he visited the Free Enterprise Uranium Mine property located at Boulder, Montana.

That affiant noted a reaction and a change in his physical condition the second trip into the mine workings of the Free Enterprise uranium property, and that following six visits to the Free Enterprise, he now reports his present physical condition — pain attending the rheumatoid arthritis has been eliminated entirely, soreness has been eliminated; swelling in hands and other joints receded, body is now more limber. Affiant further states that prior to the mine visits, he was unable to remain in bed with any degree of comfort more than two hours, and can now remain in bed ten to twelve hours and enjoy that rest period.

Affiant further states that flexibility of hands and fingers has materially increased and he can now close his hands, which he was previously unable to do. That affiant can now walk with perfect ease.

Affiant further states that based on his own observation relative to rapid improvements in his own physical condition, he now feels qualified to recommend the property to others afflicted in the Washington area.

M. E. H., Skykomish, Washington
— October 27, 1952

Case No. 1017: For about six weeks now I have experienced a great improvement in my condition (arthritis in my right hip). I visited your mine the latter part of June and felt better after each trip down the mine and for about a week after my return home.

The improvement was not lasting, however, and the pain returned. I was certainly surprised when a decided change came about six weeks ago and since then I have been entirely free from pain and stiffness for the first time in ten years.

If the cure is permanent, I shall be eternally grateful and if not, I will be thankful for the temporary relief.

Mrs. H. F. D., Portland, Oregon
— November 13, 1952

Case No. 1105: After three months following Free Enterprise Mine visits, my sinus has cleared 60%, pain 75% less, flexibility increased 25%.

Mr. E. J. B., New Germany, Canada
— December 1, 1952

Case No. 1205: I have had arthritis for the past seven years and suffered much pain, and had a stiff arm. Now I feel wonderful since going to the Free Enterprise Mine. Reporting after three months, I have no pain now.

Mrs. M. M., Gladewater, Texas

Case No. 3516: I have suffered from arthritis and asthma for many years, also spent many dollars trying to get some relief.

About a year ago I heard about the Free Enterprise Mine in Boulder, Montana, helping so many people with arthritis and asthma.

The first of last July I visited the Free Enterprise Mine. I felt terrible pain most of the time, and had to use a crutch to get around. Could not close my hands which were swollen and sore.

Due to so much pain I could not sleep. I

71

could not put my sock and shoe on my left foot or take them off without help.

After my first visit to the mine I slept like a baby. After my third visit I had very little pain. I was much happier as I could move around without all the pain. On my last visit I walked from the mine to the car without the crutch. It has been nine months since my visit to the mine and I have not had an asthma attack since last July, and I have no pain from the arthritis. My hands are normal again.

I have roofed our home (by myself), climbing up and down ladders, carrying roofing and tools. I also topped a large tree from an eighteen foot ladder. I drive our car now. All these things I could not do before I visited the Free Enterprise. I thank God for the help I have received and I intend to visit the mine again in June.

Mr. O. W. J., Clarkston, Washington
—April 11, 1953

Case No. 4132: First of all, I want to tell you of the wonderful results that I experienced from my visit to your Free Enterprise Mine. Over a period of two years, I had been taking from 12 to 50 aspirin a day, trying to shut off the terrific pain in my spine and ribs. I had made five trips to the Mayo Clinic and two trips to the Santa Fe Hospital at Topeka, Kansas. None of the doctors were able to shut off that pain, and for one year before my trip to your mine, I had been wearing a steel brace on my spine; without the brace I could not stand up straight.

After my first visit to the mine I was able to stand up straight, without the brace, and I haven't had it on since. I have taken no aspirin or pain pill of any kind since the 27th of June and can *truthfully say that I haven't even had any pain of any kind. To me it is a wonderful thing, just to feel as though I belonged to the human race again. I am as free of movement now as I have ever been, and life just seems to have a different meaning.*

I have had to spend a great deal of my time, since my return, telling people who come to see me, about my trip and the benefits that I received from my visits. I know that there have been quite a few of these people make a trip out there and I have talked with them since their return and know that they have had good results, too.

My work here has been very strenuous since my return and believe me, I would never have been able to keep up with my work if I had not received wonderful results and relief.

Again, please accept my thanks for the many courtesies and services that you people rendered me, I remain,

Very sincerely, your servant,

Mr. E. E. K., Fort Madison, Iowa

—August 17, 1953

Case No. 4465: In regards to my trip to the mine for asthma I feel just swell, and putting on weight, can eat anything and anytime now, and feel pretty peppy for a man 66 years young. I hope the damp foggy weather this winter doesn't bother me. But if it does I will be out there in the spring as I have great faith in the Free Enterprise Mine. I saw some marvelous results while there, arthritic men and women who were so badly crippled with arthritis and rheumatism, how they came in wheel chairs, and in 6 to 8 hours total mine time exposure, walked out. I will always sing praises for the mine.

Mr. J. T. J., Chicago 51, Illinois
—November 3, 1953

Note: The above writer arrived at the Free Enterprise Radon Mine suffering from a severe case of asthma. He was an employee of the C. M. & St. P. R. R. Co. He further reported "no pain or choking since my trips into the mine."

On March 7, 1953, the author spoke before the Montana Section of the Institute of Mining and Metallurgical Engineers, at the Montana Club, Helena, on a subject entitled "Importance of Uranium Transmutation Elements." Present was an industrialist from the British Isles. This English gentleman spoke, in summary, as follows:

Case No. 4528: I am not on the program tonight but wish briefly to express my appreciation for the obvious benefits I have received by reason of visiting the Free Enterprise Uranium Mine.

As executive head of a large industrial firm in England, I worked under conditions of the "blitz" and bombardment by the German fliers during World War II. Following this ordeal I became seriously afflicted with rheumatoid arthritis, and suffered increasingly over a 7 year period.

Hearing of the Free Enterprise Uranium-Radon Mine, I flew from Bermuda to New York, thence to Boulder, Montana.

During my daily visits to the Free Enterprise Mine, prior to tonight, I became free from pain, and now find, due to greater mobility of body members, that I can now engage in physical activities not previously possible. My mine visits extended from March 1st through March 7th, and these improvements were all noticeable during that time period. I was invited to this meeting by the officers of the Free Enterprise Mine operation, as stated before. Am not on the evening's program but I do wish at this occasion to express my appreciation for benefits obviously resulting from visits in the underground workings of the Free Enterprise Mine.

Mr. E. H. L., Warwickshire, England
Case No. 4565: *My rheumatoid arthritis symptoms appeared in 1937. From that time on I became gradually worse until August, 1952, I was confined to a wheel chair. I was even unable to stand. All joints of my body were afflicted, with attendant pain and swelling.*

My first visits to the Free Enterprise Mine were around August 10th, 1952. I received no startling immediate benefits such as have been claimed by other mine visitors. Rather my improvement was gradual in nature, this improvement being noticeable over a month at a time, not in a matter of days.

I made a second series of one-hour mine visits about November 2, 1953, followed by continued improvement. Now, in March, 1954, I have returned for a third series of mine visits. I am, as you observe now, out of a wheel chair, living normally, joint swelling has receded and I experience very little pain at any time.

Mr. J. R. M., Calgary, Canada
Case No. 4580: *It is with a great sense of appreciation that I relate what my visit to the "Enterprise Uranium Mine" has done for me. My arthritic condition in the spine, neck and shoulders had gradually advanced; and I was unable to obtain any relief from doctors, mineral baths and various health resorts.*

However, after six visits to the original Uranium Mine in Boulder in August, 1953, I started to show improvement and when I returned to my home in Portland, Oregon, I ceased taking any medication. The shoulder and neck condition showed the most marked improvements, and one month later I was able to take a 1500 mile auto trip to southern California, the first driving I had done in almost two years.

Some years ago many doctors hailed cortisone as a boon to arthritics. However, in quite a number of cases they were disappointed because of the detrimental side effects. Now, we who have open-mindedly investigated the Uranium Mine, learn that it is the inhalation of the radon gas which activates a lazy or impaired glandular system, thus assisting nature in the manufacture of her own body hormones.

It is a wonderful thing to learn about the latest scientific discoveries of this modern age, but how much more wonderful to be able to accept the availability of any specific discovery which directly solves one's own particular problem. I have always been interested in any scientific advance, having majored in Science, so when I visited Boulder, Montana, I was delighted to learn that the person responsible for this discovery is a sincere and scientific minded person evidently desiring to spend the remainder of his life in research work pertaining to radioactivity, and its direct benefits to mankind. More power to such a courageous and pioneering soul! Indeed, it is a great pleasure for me to tell others of the Free Enterprise Mine.

G. B. M., Cathedral City, California
—May 5, 1954

The foregoing early testimonials are not acceptable to the medical profession as evidence of benefits claimed by mine visitors. Even ten thousand testimonials would not be.

However, since receiving early testimonials, thousands of others have visited the Free Enterprise Radon Mine, many of whom have been checked for sedimentation rate, with other tests by medical doctors, and such laboratory evidence is, of course, quite acceptable, to indicate effectiveness of radon and its transmuted elements. Sedimentation rates often drop precipitously after a series of mine visits.

Following are a few of the condensed comments of visitors over recent years, data having been supplied on printed inquiry form supplied by a St. Paul medical doctor.

Case No. 4590: Arthritis: Two months after visiting, joint and muscle pains all gone. Improvement 99%.

Mr. E. D., Eureka, California

Case No. 4596: Arthritis and sinusitis: Visited since 1956. Complete relief from both afflictions, not on cortisone or other medication.

Case No. 5010: Arthritis: Duration, 5 years. Nature's radon treatment "Reactivates the body in general." 90% pain free. Am 86, feel 65. Resumed long walks.

G. M. J., M. D., Kansas City, Missouri

Case 5032: Asthma: Complete relief from asthma a week after a series of Free Enterprise Radon Mine visits.

H. H. K., Coaldale, Alberta, Canada

Case No. 5048: Arthritis and kidney problems: "Cured of arthritis, complete relief from kidney trouble."

S. C. B., Marysville, Washington

Case No. 5076: Arthritis: Duration: 15 years. "Since second day at Free Enterprise Mine I have had no arthritic pain of any kind; joint swelling gone, sleep well.

L. D. S., Eureka, California
Case No. 5091: *Arthritis: Duration: 6 years.* "All pain in hands and legs completely gone by end of treatment of twice a day for 2 weeks. At first a skeptic, I am now back to work."

Mrs. N. H., Sherwood, Oregon
Case No. 6007: *Sinusitis and thyroid gland: Sinusitis cleared. Thyroid relief:* "Stopped taking thyroid tablets which I had taken for 6 years. Circulation much improved."

Mrs. D. S., Fort Morgan, Colorado
Case No. 6021: *Arthritis: Crippled 4-1/2 years; tried much medication.* "Pain and swelling in hands, knees and feet completely gone."

I. V., Lethbridge, Alberta, Canada
Case No. 6034: *Arthritis: 15 years duration.* "Pain and swelling eliminated. Could not comb my hair. Now painting kitchen ceiling. Two M.D.s had told me I would be in a wheel chair." Reporting 8 months after visiting.

Mrs. B. M. F., Willows, California
Case No. 6052: *Arthritis: 10 years duration.* "Feel 100% benefited; no more pains or aches. Am 70 years old, feel like 20; no more medication."

Mr. F. W., Carlston. Alberta, Canada
Case No. 6122: *Arthritis, sinusitis, diabetes: Duration: arthritis, 25 years,* "Pain, swelling eliminated. No recurrence of sinusitis or diabetes."

Case No. 7015: *Skin allergies:* "Had skin allergies all my life. Eliminated 6 weeks after 29 one-hour visits to Free Enterprise."

Case No. 8044: *A recent mine visitor is quoted in full from his communication to the Editor of the* Boulder Monitor.

To The Editor:

To any sufferer from rheumatoid arthritis I earnestly recommend visiting the Free Enterprise Uranium Mine at Boulder, Montana. It has been in continuous operation every day over the past 12 years.

After going everywhere and trying nearly everything in nostrums I recently heard about the Free Enterprise Mine, and flew from my home in Detroit, Michigan, arriving here November 5th, 1963. I descended by Otis elevator to the 85-foot level of the mine to breathe the radon laden atmosphere. Incidentally, the mine does not utilize external radiation from its uranium ores, rather the elements that are transmuted from the gas, radon. I have been here for 10 days now, and can report that I am pain free, swelling has subsided in my hands, and other body members. I am out of pain for the first time in years and able to sleep soundly by reason of being free from pain.

Over a period of two months I recently spent some $1400.00 attending a clinic in Missouri. After two or three days at the Boulder Radon Mine I was able to claim more benefit by the expenditure of $18.00 than I had by spending $1400.00 in Missouri.

Sincerely, Daniel Mercan

The following case was likewise reported in the *Boulder* (Montana) *Monitor:*

Case No. 8072: Evonne Sabol, 13-year-old daughter of Mr. and Mrs. John J. Sabol of East Helena, Montana, afflicted with arthritis, is rapidly recovering by reason of visiting the Free Enterprise Radon Mine at Boulder, Montana.

Crippled, and arriving at the mine December 11, 1962, she was carried from an automobile to a reception room wheel chair. Rheumatoid arthritis had developed following an attack of rheumatic fever 14 months ago. Progress of the arthritis was rapid, causing deformation of her feet to such an extent that she was walking on the sides, not bottoms of her feet.

After 7 visits to the Free Enterprise Mine, between December 11th, 1962, and January 3rd, 1963, Evonne's improvement was estimated at 50%; she left the wheel chair and was walking with a cane. After 11 visits she discarded her cane, her ankles straightened, joint swelling receded and she now walks on the bottoms of her feet. Pain was eliminated.

Previous to the Free Enterprise Radon Mine visits, Evonne Sabol and her parents had consulted with many medical authorities, one suggesting corrective operations on her feet.

While Evonne Sabol's case appears as a "miracle cure", here is only one of hundreds of cases that have responded dramatically over the past 12 years of continuous mine operation. It is unfortunate that a number of national magazines as well as certain national organizations, presumably committed to aiding arthritic cases spending annually millions on treating arthritis symptoms, do not investigate the obvious benefits that accrue from radon exposure and its accompanying important factor of ionization. For this reason Radon Research Foundation, a Montana non-profit organization with offices at Boulder, Montana, claims that "millions of people remain unnecessarily in pain."

Here again, it is demonstrated that nature's radon therapy effects relief or cure in many cases, by reaching a basic cause of rheumatoid arthritis, reactivating the body's endocrine gland system in its production of ACTH, hydrocortisone and other normal body hormones.

The foregoing letters, statements, and testimonials are only a few from hundreds reciting the experiences of persons from many parts of the world, visiting the Free Enterprise Mine for relief from arthritis and other afflictions. The statements are sincere, written in good faith, and express deep appreciation for relief from constant pain and suffering. The home town doctors attending these cases appear not less surprised than the mine visitors themselves.

The prospective mine visitor often asks what afflictions are remedied by visiting the Free Enterprise Mine for radon exposure.

In general, visitors claim benefits for most glandular connected ailments. These include arthritis, bursitis, asthma, sinusitis, eczema, allergies and other skin afflictions. Many persons after acquiring a sense of well-being, claim improvement in sight and hearing.

Exposure to the mild radiation from breathing transmuted elements from radon, a gas, coupled with attending ionization, represents a scientific break-through, offering a remedy reaching a principle *cause* of arthritis and allied glandular afflictions. Reaching the *cause*, the symptoms of pain and swelling of afflicted joints disappear, as attested by reports of thousands of mine visitors, over a 12-year period of radon mine operation.

While individual statements of visitors are important, the statements of medical doctors who have visited and investigated the Free Enterprise Radon Mine are perhaps more

important and are included in the chapter to
follow.

CHAPTER 6

WHAT THE DOCTORS SAY

From the day radon gas was discovered in the mine workings of the Free Enterprise property, doctors, physicists, and other technicians were invited to visit the mine and observe first hand the results and claims of the scores of afflicted people visiting the mine daily. Many responded and ventured comments, others reserved opinions pending further study and evaluation.

What the Doctors Say is best expressed in an article entitled *Nature's Own Remedy* by Harold R. Tregilgas, M.D., F.A.C.S. (in collaboration on research with Wade V. Lewis, B.S., Geologist). The article appeared in *Let's Live* Magazine, published at 1133 N. Vermont Avenue, Los Angeles, California. Copyrighted in 1962, by Godfrey Thomas Publishing Company, the author has Kay K. Thomas, Editor and

Publisher, to thank for her graciousness in granting permission to reproduce the article, which follows:

For ten years the Free Enterprise Mine at Boulder, Montana, has been in continuous operation. Initially it was the first commercial uranium ore producer of Montana. It is now operated as a means of affording relief to those afflicted with arthritis, bursitis, sinusitis, neuritis, asthma, and other chronic, painful conditions which the medical profession is often unable to clear up with regular treatment.

As a medical doctor, my conclusions regarding benefits accruing to the mine visitor are based on personal observation and experience, as well as upon reports from other sources. In 1959, I developed a painful stiffness and swelling in both hands, right arm, shoulders and neck. This condition increased in severity, making it difficult to continue surgery.

I visited the Free Enterprise Mine, taking, over a 10-day period, about 30 hours of exposure to radon and its transmuted solid elements. Within weeks I experienced elimination of pain, stiffness, and swelling of body members. I have had no signs or symptoms of arthritis since, and have taken no aspirin for pain for the past two years.

Since visiting the Free Enterprise Radon Mine I have sent many cases there, a high percentage of which responded to the treatment.

Results attained at the Free Enterprise Mine, especially for rheumatoid arthritis, are not without precedent. In 1934, the beneficial effect of radon, present in more meager amounts in springs and spas, was discussed by Francis J. Scully, M. D., in a bulletin entitled "The Role of Radioactivity of

*Natural Spring Waters as a Therapeutic Agent."
reprinted from the* Journal of the Arkansas Medical Society, Vol. XXX, *March 1934, pp. 206-214.
Dr. Scully emphasizes that radon can be employed
for a long period of time without any adverse
after-effects and states that arthritis in all forms
benefits from this type of treatment.*

*A parallel to the Free Enterprise Radon Mine
operation is found at Bad Gastein, Austria, where
a 1,500 meter adit affords a copious amount of
radon gas. German pamphlets reporting on European arthritics show clinical evaluations virtually
identical to observations at the Montana radon
mine.*

*The Bad Gastein mine is described in the
Journal of* the American Medical Association,
*June 30, 1956, Vol. 161, p. 917, under MEDICAL
LITERATURE ABSTRACTS, in an article entitled "Combined Treatment by Radium Emanation and Hyperthermy of Bad Gastein," by Otto
Henn. The* Journal *article, in referring to radon
gas, concludes:*

*"It exerts an influence on the autonomic nervous system, improves the circulatory state, and
causes removal of waste from the organism and an
activation of hormone producing organs, particularly of the pituitary-adrenal system. Indications
and contraindications for this type of treatment
are apparently the same as those for corticotropin
(ACTH) and cortisone therapy."*

*Evidently two principal environmental factors
in nature account for the physical improvements
now claimed by those thousands who have visited
the Free Enterprise Mine: (1) radon therapy, the
utilization of radon's transmuted elements, producing mild but effective internal radiation reaching*

the blood stream, the body cells, and the endocrine gland system; (2) ionization, a product of radon's radiation, reported as stimulating defense cells of the body, inducing better utilization of oxygen.

A recent paper entitled "Medical Hydrology," by Igho Hart Kornbleuh, M.D., and Paul K. Kuroda, Ph.D., states:

"Studies of the biologic influence of unipolarly ionized air show some striking similarities to the effects of radium emanation (radon). The results so far achieved indicate that inhalation of this gas or of ionized air is by far most effective.... Clinical evaluations have established the following effects of radon:

"A pain controlling quality, stimulation of the inner secretory glands, increased diuresis and excretion of uric acid as its most conspicuous properties.

"It appears that radon in carefully controlled quantities exerts a rather specific eubiotic influence without any undesirable after effects."

According to an article entitled "Ions Can Do Strange Things to You," written by Robert O'Brien and appearing in the Reader's Digest of October, 1960, experiments are going forward at Northeastern Hospital in Philadelphia, using artificially developed ionization for post surgical pain, and particularly for severe burn cases. The article states:

"Negative ions in the blood stream accelerate delivery of oxygen to our cells and tissues . . . Researchers also believe that negative ions may stimulate the reticulo-endothelial system, a group of defense cells which marshal our resistance to disease."

Other sources claim that the internal radiation acquired at the Free Enterprise Mine *stimulates the boss pituitary gland in its production of ACTH, the latter acting upon the adrenal cortex in its production of hydrocortisone, the body pain killer.*

In October, 1961, an International Conference on Ionization was held at the Franklin Institute, Philadelphia, sponsored by the American Institute of Medical Climatology. Medical doctors, as well as those from other professions, attended from France, Germany, Sweden, Denmark, Russia, and the United States. Proceedings of this meeting emphasized the importance of ionization for body well-being.

It is evident that physical benefits observed at the Free Enterprise Mine are due to induced body chemical changes, increased body hormone production, and are certainly not of psychosomatic origin, except to the extent normally anticipated by any kind of therapy.

The subject of radon gas, its transmuted elements and attending ionization, certainly merits further research, under careful medical supervision. Such studies should be very rewarding and worthwhile.

James M. Wolfe, M.D., now of Tahlequah, Oklahoma, after visiting the Free Enterprise Radon Mine, authorized the following statement, published in the *Boulder Monitor.*

"I have been at the Free Enterprise Mine for seven days, in the capacity of an investigator. I have talked to several dozens of patients whose ailments have varied greatly. I have seen and carefully noted excellent results in rheumatoid arthritis, sinusitis, asthma and migraine.

"I have read records of diabetics having left here with no need of insulin.

"*This radon therapy is so new, the results it gives so amazing, the possibilities so great, that its ultimate is impossible to visualize.*"

Another doctor, George M. Jaquiss, M.D., physician and surgeon of Kansas City, Missouri, holding Medical and Bachelor of Arts degrees, 58 years in practice, had this to say, as published likewise in the *Boulder Monitor*.

"*While visiting the Free Enterprise Mine, I have observed and talked with different people from many states and Canada, who have received wonderful benefits by reason of radon exposure. As for myself, have been here only one week and I have received much benefit for arthritis and, from my studies of the internal radiation and reactions, I expect to continue to improve, and be back to normal in a very short time.*

"*Since 1938, I have had much hospital and clinical experience treating different diseases and I find in the arthritic cases that the usual treatment of arthritis does not accomplish much for me, but I believe the natural radioisotopes found in the Free Enterprise Mine atmosphere are doing the work that the methods I have been using have failed to accomplish.*"

One of the most astounding finds in research is a Bulletin entitled "The Role of Radioactivity of Natural Spring Waters as a Therapeutic Agent," by Francis J. Scully, M.D., Hot Springs National Park, Arkansas. It is a reprint from the *Journal of the Arkansas Medical Society*, Volume XXX, March, 1934, pp. 206-214. Dr. Scully's observations and conclusions are remarkable in that they were made over 30 years ago. Excerpts follow:

EFFECTS OF RADIUM RADIATION (RADON)

*Observation has shown that radium emana-
tion, radon, has a powerful influence on the human body. It has also been noted that the physiological effects that it produced are quite varied.
Certain tissues of the body seem to be more strongly
affected than others. These are particularly the
blood making organs, the Lymphatic tissues, the
ductless glands, the liver, the kidneys and the
brain. It is upon the cells of these tissues and
organs that the strongest influence is exerted.*

*Radium emanation, radon, activates cell func-
tion and facilitates the elimination of waste products and toxins. There is an activation or increase
in the intensity of the vital energy of the cells. We
know that the vital energy of growth and repair is
most active in young children and that it gradually diminishes as an individual grows older. It
has been found that radium emanation has the
effect of strengthening or fostering this vital energy....*

*The stimulating action of emanation, radon,
in biochemical processes is shown especially in the
increased growth of young plants exposed to its
action.... It is also manifested by the stimulation of
the blood making tissues, and the sex glands, and
an increase in the heart action and basal metabolism. These effects are frequently accompanied by
an increased feeling of strength and rejuvenation.
Small doses have a stimulating effect on the sex
glands. Data has found it beneficial in diminished
potency.*

*The influence upon metabolism or growth in
the body is also very decided, especially in the
metabolism of urea promoted, and the excretion of*

uric acid increased. This is probably due to an activation of the chemical processes in the body which oxidize the uric acid salts and aid in their elimination. There is also an increase in the activity of the kidney....

Since we have seen that radium emanation has a powerful effect, its use would be indicated in certain types of bodily disorders. The conditions most benefited are those disturbances due to toxemias either bacterial or metabolic in origin. This includes such conditions as chronic malaria, chronic arthritis or rheumatic joint disorders, chronic nephritis, chronic skin diseases. Arthritis in all forms does well with this type of treatment.

It is surprising and much to the credit of Dr. Francis J. Scully that he was able to reach scientific conclusions to the extent he did over 30 years ago; particularly when one considers the Arkansas radon source occurred only in modicum amount in spring waters compared to the copious amount of radon determined in the underground atmosphere of the Free Enterprise Mine workings. To the author it appears a medical tragedy that the importance of Dr. Scully's findings were not recognized in 1934, that facts determined by him were not investigated for the benefit of humanity. His truths were, in effect, buried for over 30 years.

Over the past 12 year period of continuous operation of the Free Enterprise Radon Mine, many other medical doctors have cooperated in investigation and research on this scientific subject, but none of them is quoted unless their statements have been published or authorized for publication. Correspondence from doctors is always invited and respected.

The author has never been faced with any direct opposition from any organized medical group. He has never seen other than favorable reviews on the first edition of *Arthritis and Radioactivity.* However, it is coincidental that 12 years of opposition have attended several therapy discoveries asserted beneficial to humanity.

Alexander Fleming, discoverer of penicillin, is quoted as saying:

"Penicillin sat on the shelf for 12 years while I was called a quack.

I can only think of the thousands who died needlessly because my peers would not use my discovery."

Now, Andrew C. Ivy, M.D., Ph.D., of the Roosevelt University, Chicago, sponsor of Krebiozen, a product employed in cancer cases, is reported as "at war" with organized medicine and presently censored for his activities in its use and application to some hundreds or thousands of cases. His period of opposition is likewise 12 years.

Dr. Ivy is high in his profession, has taught university classes for over 45 years, and is recognized for his ability in research. The author is not qualified to comment on the merit of Krebiozen, the controversial drug, but it does appear that every effort should be made to exhaust all directions of research in determining the possibilities of Krebiozen as a remedy or cure, especially when sponsored by a man of the high caliber of Dr. Andrew C. Ivy.

The subject of radon therapy and ionization, factors utilized by the Free Enterprise Radon Mine at Boulder, Montana, although criticized by uninformed sources over a similar 12 year period, does not have the problems other forms of new therapy have experienced. Too many individual medical doctors, both in Europe and the United States, accept this subject as an effective, basic and beneficial therapy as supplied by nature. They recognize it as one therapy that reaches a basic cause of arthritis, not one that treats the symptoms only. It does appear tragic, however, that due to periodic uninformed criticism, thousands of afflicted people should continue to suffer needlessly in pain.

Three phases of radiation are already in use by the medical profession:

1. *X-ray*

2. *Radium Radiation*

3. *Artificial radioisotopes, such as Cobalt 60, Iodine 131, and others*

The doctors list the above three phases of radiation in their radio-therapeutic armamentarium, while no study to speak of has been given to a fourth phase: natural radioisotopes, including radon and the transmuted elements which follow radon as solids in the uranium decay series. The first three represent largely external radiation, while the phase being used at the Free Enterprise results in a modicum of internal radiation, safe but vital in effecting health benefits.

No tuberculosis case, based on advice of specialist, is permitted to visit the Free Enterprise Mine.

CHAPTER 7

FRAUDS YOU WILL ENCOUNTER

Since discovery of radon gas in the Free Enterprise Uranium Mine workings, scores of mining properties and "radiation" centers were opened, not only in the vicinity of the Free Enterprise Mine in Jefferson County, Montana, but at points from Seattle, Washington, to Denver, Colorado, and south into Arizona and California. A few of the mines contain some radon gas. Some operators proceeded with serious intent, some operated in ignorance, others simply did not care as long as they collected money from an unsuspecting public. Many of the mines, now radioactive, had no evidence of radiation prior to opening.

Some promoters, not content with real mine workings, boldly constructed artificial adits or tunnels with portals abutting main highways with timbers, lagging and all, then lined such adits with purloined uranium ores. This uranium was secreted in the walls, roof and floors. Some of such promotions contained uranium bearing material from discarded waste dumps assaying as little as 0.01 percent U_3O_8 (Uranium Oxide).

Such contraptions constructed to snare the "mine visitors" contained so little uranium that any possible radon gas emanations from the ore would be inconsequential, certainly not sufficient to be in any manner beneficial.

Laws and regulations enunciated by the United States Atomic Energy Commission relating to the handling of uranium ore are clear. Quoting from *Prospecting for Uranium*, published jointly by the U.S.A.E.C. and the U.S.G.S. (1951), pages 50 and 51:

A license from the Atomic Energy Commission is needed to sell, transfer, or receive uranium and thorium ores which have been removed from the ground, no matter where or when they were mined. The procedures for obtaining this license, which are simple to follow, are included in a regulation issued by the Commission. (See appendix 4.) The ores subject to licensing — called "source materials" — are those that contain by weight 0.05 percent or more of uranium or thorium or any combination thereof. A license may be issued to authorize a single transfer or continuing transfers.

Licensing regulations do not apply to very limited movements of uranium or thorium which do not involve the transfer of control, possession, or title, to another person, firm, or corporation. This means, for example, that a company owning several mines may collect and move ore from

*those mines to a central stock pile on its own
property without obtaining a license from the
Commission.*

Thus both buyer and seller are required to
have an Atomic Energy Commission License to
handle any uranium or thorium ores exceeding
1/20th of 1 percent uranium or thorium.

While it has taken nature millions of years
to make an area radioactive, it appears that man
can perform the same function in a day — by
dubious methods.

Illustrative of proposals made available to
the public, we quote from a letter dated August
27, 1952:

*This letter is being sent to you by Special
Delivery because we believe it is important enough
to warrant it.*

*Enclosed you will find a page from the Hel-
ena paper. On pages 22-24 of Life Magazine, July
7th, you will find an interesting article entitled:
"Arthritics Seek Cure in Radio-Active Mines."*

*So that you may give your patients the Gamma
ray treatments right in your office, we will rent
you a Radio-Active Ore chest and a Geiger counter
for the sum of $250.00 per year which is less than
70 cents a day. This chest is approximately 31 x 16
x 7-1/2 inches in size. It contains radioactive ore.*

*We are prepared to deliver a limited number
of machines. If you wish to be one of the renters
please sign the agreement at the bottom of this
page. Your chest will be expressed at once.*

A "doctor's" newspaper advertisement read:
*Doctor D.C. of the . . . Clinic announces the first
and only machine to give you treatments from
Radioactive Ores.*

A *"Health Center" using uranium ores*

advertised: Uranium treatments are not expensive and you are not being "taken in by sharpies."

At a "Uranium Center" near Spokane, Washington, the *Spokesman Review* reported, "Customers Pay to Sit on Radioactive Ores" which "Uranium Ores" turn out to be low grade thorium ores.

A "Supply Company" asserted: "NOW! YOU MAY TEST THE HEALING effect of the Gamma Rays of URANIUM ORES In Your Own Home!"

Over 100,000 folks of all ages and all walks of life have visited the famous mines at Boulder, Montana.

Thousands of these people suffering the painful effects of crippling arthritis, rheumatism, muscular ailments, calcium stiffened joints, and the miserable feelings of hay fever, sinus trouble, eczema, etc., have come away from these visits claiming to be benefited greatly or to a lesser degree. Thousands have made numerous trips and are enthusiastically planning additional excursions for further relief.

On the theory that there were many of these who could use the healing powers of radon gas, The _____ Supply Co. introduced and marketed uranium minerals in five pound containers. These sacks have now been shipped to all points of the country, and the demand grows daily. Persons using these minerals have found that by placing them in an oven and placing the heated package beside the affected part, the benefits are greater.

A Seattle, Washington "Health Center" referring indirectly to the Free Enterprise Mine, states: "No longer will it be necessary to travel hundreds of miles to stand in line on the hillside to avail yourself of the possible advantages that may be yours."

On February 25, 1953, the following article appeared in a Washington paper under the caption: "CURE-ALL ROCK ORDERED SEIZED."

Radioactive minerals advertised as cure-alls for many types of ailments drew a judicial frown in federal court yesterday. Federal Judge Sam M. Driver ordered seizure of 79 bags of radioactive minerals imported to a Spokane home from Boulder, Montana, and being marketed here for the cure of various ailments.

A libel action asking seizure of the material was filed by the United States attorney's office in Spokane at the request of Inspector Horace A. Allen of the Federal Food and Drug Administration. Allen said the radioactivity was of "low potency and not dangerous."

The government claimed the material was misbranded, in that it purported to be an "effective treatment for various forms of arthritis, rheumatism, muscular ailments, calcium-stiffened joints, hay fever, sinus trouble, eczema and other unusual diseases and disorders." It also charged the radioactive material was represented as being capable of restoring and rejuvenating cells and glands. It is charged that all these claims were false and misleading.

The government said the bags of radioactive minerals were accompanied by about 1000 leaflets stating, "Now you may test the healing effects of the gamma rays or uranium ores in your own home."

The above is quoted to indicate the scope of activities that developed around the original radon discovery at the Free Enterprise Mine.

Every questionable project has had its adverse effect on legitimate operations, for the reason that radiation covers relatively a new science, and it is difficult for the public to differentiate between a sincere, scientific approach to a new subject, and dubious promotional methods.

A Portland, Oregon, "uranium ore room" opening to the public the summer of 1953 advertised:

WE OFFER NEW HOPE to those of you who suffer from arthritis. Exposure to the rays from radioactive material, an ore that comes from a mine in Montana, greatly relieved many persons. We have brought a large quantity of this radioactive ore to our newly opened Arthritis Center for your convenience.

Sorry, no phone. We opened sooner than telephone installed. Come in and talk to us. Maybe you will be among those who fortunately get such blessed relief from pain and suffering of arthritis through controlled exposure to this radioactive mineral ore.

Still another news story in a Spokane, Washington, paper related under the title, "Customers Pay to Sit on Radioactive Ores":

A recently opened "Uranium Center" on the highway just north of Spokane caters to persons seeking relief from various ailments. Shown at the left are two unidentified "patients" seated on a bench which is filled with radioactive ores. The man has his foot on a box filled with ores. On the right a chunk of ore rests on the table beside the manager. He is holding a small Geiger counter to demonstrate the ore's radioactive properties. The manager asserted his organization makes no claims as to the ore's healing powers, but that patrons *have claimed relief. He said the ore — about 6000 to 7000 pounds of it — was imported from the Last Chance mine in southwestern Montana and that customers pay $3.00 an hour to sit in the place. The federal food and drug inspector in Spokane, says he isn't sure of his agency's authority in the matter but is "reporting to Washington." The manager says he opened up for business only last Wednesday, and the first patients are those who have patronized the Boulder, Montana, mines for similar purposes or know people who have done so. Government geologists, he said, have permitted removal of the ore because it is not harmful.*

Note: The so-called uranium center ore turned out to be thorium ore from the Montana mine named.

Other "uranium room" operators claimed that "during your visit you will be subject to bombardment of Gamma rays from all sides and within lung area, that thousands of your body cells will be ionized but your body will not become radioactive."

One mine owner in brochure states:

One of the most interesting cases after the mine treatment was that of a 15-year-old dog. She had been crippled with rheumatism for several years and could scarcely be persuaded to get up for meals. She was a family pet and her owners hated to have her destroyed. They took her down into the mine a few times and she began to perk up. After a few treatments she was running up and down the hills like a puppy.

Another version reported the crippled dog saw a rabbit on reaching the surface, gave chase, and the dog hasn't been seen since.

One mine owner in the Butte-Helena area observed that his claim area in the vicinity of his mine portal was extremely radioactive. Immediately, he opened as a "radioactive health mine."

In this instance the radioactivity was caused by contamination of a wide area across western Montana when radioactive products settled over several counties by reason of radioactive particles carried northward from site of a Nevada atomic bomb explosion. Products from the explosion, carried northward on air currents, made a tremendous ground surface area intensely radioactive. This extended to trees, roof tops, vegetation and bodies of water. After opening his mine the owner undoubtedly realized the real reason for temporary radioactivity on his claim, but by that time, advertising was out, mine visitors were arriving by the scores, dollars were being collected for mine visiting, and so, regardless of merit, the property continued in business.

These are a few examples of what the uninitiated mine visitor faced on arrival in Montana from out-of-state. Cases are known of people traveling 2500 miles to visit a radon property, arriving in pain, leaving in pain, and continuing in pain, having visited an ineffective mine property, without having experienced even the least chance in the world of receiving relief from affliction. Such people spent hundreds of dollars in travel from distant states, returning home in tragic disappointment.

It is not claimed that all properties discussed in this chapter are frauds, intentional hoaxes. But it has become obvious since the original discovery of the Free Enterprise Mine that no attempt was made, at most of the properties, to determine scientific aspects relating to each of them.

At last count 18 old abandoned mine workings had been renovated along or near Butte-Helena Highway 91, and opened to the unsuspecting public. Few contained appreciable natural radioactivity, and instances are reported where mine workings have been made artificially radioactive, simply by "salting" mine workings with extremely low grade uranium material from waste dumps, in a manner similar to that used in salting a "gold mine" in the early days.

A few mines besides the Free Enterprise do contain some radon gas within their underground workings. Where this fact has been established in any Montana mine the owners should have written evidence to indicate the amount of radiation in micromicrocuries present per liter of mine air and radon gas to substantiate such claims. Those mines that are bona fide in this respect are readily identifiable.

Some travelers visit the Free Enterprise Radon Mine for an initial single one-hour mine visit, then through persuasion decide to go to another mine, for several successive days. About the second or third day after the single Free Enterprise visit the traveler claims a beneficial reaction, and attributes relief to the second mine visited. Since favorable reaction to radon exposure appears to take place uniformly with many people within the second 24-hour period after initial visiting, it is natural to observe benefits after leaving, wherever the person may be, whether sitting in a hotel room or sitting in the second mine containing little or no radiation. Such people, having breathed radon initially in a real radon producing mine, thereafter, erroneously, often attribute claimed benefits to the second or third mine visited.

Fortunately, most of the "uranium" frauds here noted have been discontinued, or closed by authorities. Some, intentional or unintentional hoaxes, were voluntarily discontinued because afflicted visitors claimed no beneficial response. Others were closed, after investigation, through legal action. Conversely, the authentic Free Enterprise Radon Mine has remained open every day, over a decade, and now has a national and international reputation for positive results achieved. Some mine visitors have returned to the Free Enterprise at intervals over the past 12 year period for additional benefits, or simply, as they state, as a means of "health insurance."

An article entitled "How to Spot a Quack," appearing in *Parade Magazine,* newspaper syndicated, in October 1963, and attributed to a medical doctor, the then president of a medical organization, stated in text, "Why would a man crippled with arthritis pay by the hour to sit in an abandoned uranium mine?"

The foregoing statement, of course, infers that uranium mines, used for health purposes, particularly for arthritis, are frauds. The good doctor is quite correct, in considerable measure, since many of these so-called mines, have, in years past, been "salted" with radioactive materials, alleging benefits that could not indeed be attained from the external radiation from the radium contained in uranium ores. However, it is unfortunate that the doctor-author did not supplement his statement by noting that the Free Enterprise Mine, at Boulder, Montana, is quite an exception to the rule.

A man crippled with arthritis sits in the authentic Free Enterprise Mine, now in its twelfth year of continuous operation, over a series of visits for the following reasons:

1. He sits there by reason of the presence of radon gas contained at the 85-foot level of the Free Enterprise Mine.

2. He sits there because, after twelve years of observation and research, it is well established, by scientific investigation, that by breathing and exhaling a controlled amount of radon, his endocrine gland system, through radon's transmuted elements, and the attending factor of ionization, is stimulated toward normalcy in the production of body hormones.

3. It is now well established that radon's transmuted elements produce a mild, but effective, internal radiation, reaching the blood stream, the body cells, and the endocrine gland system, inducing, through direct or catalytic action, the pituitary to produce ACTH, the adrenal cortex to produce hydrocortisone.

4. He sits in the mine because radon represents a beneficial ionizing influence, that is reported to stimulate the defense cells of the

body, inducing better utilization of oxygen.

5. He sits there because clinical evaluations have established the following effect of radon: a pain controlling quality, stimulation of the inner secretory glands, increased diuresis, and excretion of uric acid, these being some of its most conspicuous properties.

6. He sits there because radon, in carefully controlled quantity, exerts a specific eubiotic influence upon the body chemistry, without any undesirable after-effects.

7. He sits there because it has been demonstrated that physical benefits observed at the Free Enterprise Radon Mine are due to induced body chemistry changes, increased body hormone production, and are certainly not of psychosomatic origin, except to the extent normally anticipated by any kind of therapy.

8. He sits there because research through Radon Research Foundation, a non-profit organization in Montana, indicates that *a cause of arthritis is known, and that nature has a cure for arthritis.*

9. He sits there because such research indicates that there are two basic causes of arthritis and kindred glandular connected ailments:

a. *An external cause:* Due to stress, strain or shock, of physical, mental or emotional origin, followed by:

b. *An internal cause:* Retardation of the glands' activities, resulting in non-normal production of body hormones.

10. He sits there in the Free Enterprise Radon Mine because, within a twelve year period of continuous operation, thousands of people have been benefited for arthritis, bursitis, asthma, sinusitis, eczema, and kindred ailments. He visits the mine because exposure to the mild radiation from breathing transmuted elements from radon, a gas, coupled with attending ionization, represents a scientific breakthrough offering a remedy reaching a principle *cause* of rheumatoid arthritis and allied glandular afflictions. Reaching the cause, symptoms of pain and swelling of afflicted joints disappear, as attested by reports of thousands of mine visitors.

11. More and more arthritics continue to arrive and sit in the radon mine atmosphere because medical doctors have checked many patients laboratory-wise, and found their sedimentation rates drop precipitously, while red and white cell count approach normalcy following a series of Free Enterprise Mine visits.

12. German written clinical evaluations received from doctors of Innsbruck University show parallel findings in use of radon found in a

1500 meter adit neighboring the Bad Gastein Spa, in Austria. This Austrian mine is described in the *Journal of the American Medical Association*, June 30, 1956, Vol. 161, No. 9, page 917, under "Medical Literature Abstracts," in an article by Dr. Otto Henn. The Journal article, in discussing radon, concludes: "It exerts an influence on the autonomic system, improves the circulatory state, and causes removal of waste from the organism, and an activation of hormone producing organs, particularly of the pituitary-adrenal system. Indications and contra-indications for this type of treatment are apparently the same as those for corticotropin (ACTH) and cortisone therapy."

Officers and directors of the Radon Research Foundation feel that the doctor above indicated did not mean to infer that the Free Enterprise Radon Mine was other than an authentic operation, since many individual doctors, members of the American Medical Association, are now sending some of their patients to the Free Enterprise Radon Mine, or are arriving at intervals for their own afflictions. The Radon Research Foundation, and the company operating the Free Enterprise Radon Mine are in full accord with the doctor's views relating to those operations fraudulent in character. In fact, nine years ago, Wade V. Lewis, in the first edition of this book included a chapter relating to those fraudulent operations that were endeavoring to imitate the original and authentic operation, known as the Free Enterprise Radon Mine. Incidentally, the Free Enterprise

Radon Mine, the first shipper of commercial uranium ores from Montana, is not an "abandoned" mine and still contains commercial uranium orebodies, but is utilizing other elements of the sixteen elements known in the uranium transmutation series, including radon, for the physical benefits of mankind. It should be noted that not one uranium mine in hundreds produces a copious amount of radon sufficient to benefit the mine visitor. This explains why the Free Enterprise Mine, at Boulder, Montana, is unique in effecting important benefits that accrue to mine visitors afflicted with glandular connected ailments.

CHAPTER 8

RADIATION SOURCES

Radiation and radioactive materials are naturally present with us under all environments. They are contained in the air, cosmic radiation from outer space, the earth's rocks and soils, surface and underground waters, sea water, even certain vegetation. Radiation is widely distributed and this accounts for the radiation background present everywhere as shown by Geiger Counter, Scintillometer, or other radiation instruments.

COSMIC RADIATION

Cosmic radiation or "cosmic rays" bombard the earth constantly from all directions. This type of radiation usually accounts for the major part of the radiation background at any particular point. When a radiation counter is near a high grade uranium orebody the cosmic background count, of course, becomes an inconsequential part of the total radiation reading. In the northeast portion of the United States 1 cosmic ray particle is reported for each square centimeter per minute. Cosmic rays are said to increase 5 times from sea level to an elevation of 15,000 feet, and 75 times at about 55,000 feet. Thus, altitude, together with other factors, appears to account for variations in cosmic radiation.

RADIATION IN ROCKS AND SOILS

Radioactivity, in some degree, exists in all soils and rocks of the earth's crust. In the soil it is due to erosional processes and weathering of rocks in place, whether of sedimentary or igneous origin. For example, when uranium occurs in certain sedimentary sandstone beds in the Utah-Colorado plateau area of the United States, such concentrations represent commercial ores, mineable at a profit. In Montana, commercial ores have recently been discovered and mined from veins traversing the monzonite, a phase rock of the granite. Outside the veins, however, the country rock contains significant amounts of uranium minerals dispersed in millions of tons of the granitic mass, covering many square miles. The gram or two of uranium contained in each ton of granite within a given area is meagre in amount but makes its contribution to the radioactivity of soils in its own locality.

It has been estimated that radioactive minerals are present in the earth's surface soil mantle in the following proportions: uranium, 6 parts per million by weight, radium, $2x10^{-6}$ parts per million by weight, and thorium, 12 parts per million by weight. While these three elements are present in meagre amounts per cubic foot of soil, it is estimated that one square mile of surface mantle, one foot thick, contains one gram of radium, three tons of uranium, and six tons of thorium.

RADIATION IN AIR AND WATER

The uranium and thorium decay series eventually, after millions of years, produce in nature two gases, radon and thoron, respectively. These are derived from the earth's rock mass and diffuse into the air. These gases are present in minute amounts everywhere, but in greater concentration in areas containing substantial amounts of uranium and thorium.

It has been ascertained that due to temperature inversion processes, radon and thoron gases, together with their decay products, collect near the ground and are not diffused in large volumes of air. Accordingly, people breathe these gases contained in the air daily, the gases varying of course in amount, depending on geographical locality.

Radioactivity in water is due to absorption of radioactive elements, including radon, from the air by rain water or circulation of ground waters. Principle minerals contained are uranium, radium, and radon or decay elements from these three minerals. Locality, again, governs content.

RADIATION IN PLANT LIFE

Radiation is found in plant life, trees and shrubs, often in appreciable, measurable amounts, where vegetation growth occurs at or near uranium ore deposits. A striking example of this is found in the recent work of a woman scientist, employed by the Geological Survey. Reasoning that the roots of desert trees and vegetation may go 50 to 75 feet below the ground surface, she assumed that commercial uranium deposits in the Colorado-Utah plateau area might be found by analyzing chemically the plant and tree growths. Such field work, as pruning a juniper tree, brought practical results in discovering new uranium ore bodies. This scientific approach follows the same plan as that used in determining minerals such as selenium found in range grasses and vanadium determined in corn stalks.

RADIATION IN THE HUMAN BODY

Because we are exposed to radiation from so many different sources, cosmic rays, air, water, soils, rocks and vegetation, such radiation is constantly reaching the human body, either internally or externally. When radiation is increased beyond normal factors for any reason, such exposures may be beneficial or harmful, dependent on the cumulative amount, the intensity, time of exposure, and kind of such radiation. It is, therefore, imperative, that all types of such radiation be at all times under a technician's or doctor's direction, whether it be applied through X-ray, radium, employment of artificial radioisotopes, or other phases of radiation.

Radioactive potassium, K^{40}, is normally present in the body in significant amounts. Potassium is present as about 0.35 percent of the weight of the human body, while only 0.012 percent of this amount is K^{40}, the radio-element. The body of a 165 pound man would therefore contain about 262 grams of potassium of which about 0.0314 grams will be K^{40}. How important this small amount of radioactive K^{40} may be to the chemistry or normal functioning of the body is not clear, although it is reported a Holland scientist years ago suggested, by reason of his studies, that a certain amount of radioactivity, extremely meagre in amount, within the human body is necessary for continued heart action.

Most certainly, the human body being the chemical laboratory that it is, the many trace elements may have important uses in normal body functioning, and, conversely, the absence of any element normal to body chemistry and activities may create imbalance.

By definition the air we breathe contains nitrogen, and oxygen, nearly in the ratio of 4 volumes to 1, with lesser amounts of argon, carbon dioxide, water vapor, and minute quantities of helium, krypton, neon and xenon. Although the trace elements are inert gases we still do not know their possible importance on the chemistry of the human body. Radon, likewise, in widely varying amounts is present at the earth surface. Its importance is becoming apparent.

A scientist, high in his particular profession, was asked what study he would undertake were he to start over again.

His reply: "I would study body chemistry.

There is so little known about it."

Knowing that radon, a gas, is normally present in the air we breathe virtually everywhere, it is reasonable to presume that nature supplied such radon for a purpose. Radon is, of course, generally present in very meagre amounts. Again, can this minute amount, with its attendant rapidly forming and changing transmutation elements, possibly be quite as necessary for body chemistry as the immensely greater amount of oxygen we utilize daily? The somewhat greater amount of gas breathed in a Montana radon mine, may, as has been suggested, supply a lack and body need, through body acquisition of those radioactive transmutation elements, or at least serve as a catalyst to effect an end result.

Many people afflicted with arthritis or other ailments, arriving at the Free Enterprise Mine, come from regions with much rainfall, long damp seasons, and localities bordering large bodies of water. It is known that radon is quite soluble in water. This being the case and radon having a short half-life of only 3.82 days, it may be supposed that radon may be nearly lacking in the atmosphere blanketing areas bordering or near large bodies of water or regions of great precipitation, and frequent rainfall. It has been suggested that rheumatoid arthritis and kindred ailments would be prevalent in such damp regions, as for example, the Florida swamps, the Great Lakes region, and the Seattle-Tacoma bay areas, and this is reported to be the case. One person has concluded that arthritis could, in some cases, be in part caused by the total absence of a normal amount of radon in particular localities.

Hence, it may be inferred that breathing a greater amount of radon benefits arthritis and kindred ailments, since nature's radioisotopes, by reason of internal radiation, do activate the body's defense system, stimulating the endocrine glands toward normal production of body hormones.

Radon, in wet regions, may be quickly eliminated from the air by being absorbed in water. While internal radiation to some degrees is doubtless acquired in the human system by drinking such water in particular localities, extreme dilution, because of large volume of water, doubtless reduces radiation contained to a negligible factor.

RADIATION FROM THE H-BOMB

Because Albert Einstein questioned accepted laws of physics, he was able to derive the formula $E = MC^2$. A simple equation, but stating that latent energy contained in a given mass is equal to the weight of that mass times the speed of light squared. Einstein's equation resulted in the H-bomb, the most devastating thermonuclear explosion in all history. Its intensity is reported reaching 250 to 750 times the power of the Hiroshima A-bomb, and about three times the total power expected. Accordingly, all previous A-bomb protection planning seems obsolete. Radiation extended for 800 miles beyond site of explosion, and pressure from the burst was felt 200 miles distant.

Contamination from radioactive particles covered thousands of square miles, proving that annihilation of one American city could mean serious secondary damage to other adjacent or distant cities.

The congested East and West coast cities would become prime targets. One H-bomb dropped by the enemy on Los Angeles could have its secondary effect on San Francisco, while one dropped on New York would affect Washington, D. C.

It is stated that the terrific temperature of the H-bomb explosions resembles that of the sun, maintained for millions of years by "fusion" of hydrogen atoms. The H-bomb burst has been rated equivalent to the power of 5,000,000 tons of T.N.T.

The advent of the H-bomb ushers in a whole new field of investigation relating to the many possible radioisotopes created by the explosion. Japanese fishermen reported being showered with "radioactive particles" resembling pulverized coral ash, with attendant burning and damaging effects.

Subsequent studies have determined the types of long half-life radioisotopes that were created by the initial H-bomb explosion and those that followed. Because of the publicity given the damaging aspects of radiation resulting from nuclear explosions, it is often necessary to assure the Free Enterprise Mine visitor that there is no similarity or possible comparison to the mild radiation encountered in the underground Free Enterprise atmosphere.

Again, it may be seen that radiation offers two extreme alternatives: (1) Intense radiation with the purpose in mind of annihilation of civilization, and (2) Proper use of relatively mild forms of radiation for the relief and benefit of afflicted mankind. Man has discovered the use of universal laws that existed before he was thought of in the process of creation. Today, man must decide whether he will utterly destroy himself, or, rather, be guided in a positive direction to constructive things by the Hand of the Great Scientist.

CHAPTER 9

URANIUM VS. RADIUM VS. RADON

Much of the criticism of the Free Enterprise Mine, coming at all times from many sources, technical and nontechnical, was due to misinformation, misunderstanding, and lack of differentiation of three separate and distinct elements, uranium, radium, and radon.

When the first news of rapid and startling "cures" as claimed by arthritics and others hit the newspaper headlines and national magazines, the changes wrought in afflicted people were attributed immediately to "psychosomatic" changes, remissions, subjective influences, or simply change to Montana Rocky Mountain altitude. Others claimed outright fraud, and use of "worked-out" mines as a means of preying on afflicted people. These assumptions will be discussed briefly, after noting the different physical and chemical characteristics of the three elements, uranium, radium, and radon.

Uranium is element 92, a metallic mineral, heavy, hard, nickel-white in color in purified state, with an atomic weight of 238.14. Its half-life is listed as 4.55 x 109 years or 4.55 billion years. Its immediate decay elements as listed in physics texts, are thorium, protactinium, a second uranium (U^{234}), and thorium (ionium). The first two decay elements have a short half-life, the latter two, a long half-life.

The next element produced in nature by transmutation or decay, the sixth in the uranium series, is radium, Ra^{226}, element 88, with a half-life of approximately 1600 years. It possesses alpha and gamma radiation. Radium is well known for its use in external radiation. Doctors employ milligram amounts of this powerful element for short time periods when enclosed in protective lead shield containers, directing the rays from one unshielded side only, upon an afflicted part, as for example, in removal of a wart. Radium decays directly to one element, radon, a gas, all other elements of the series being solids. By reason of the relatively long half-life of radium, 1600 years, the element is particularly dangerous if taken orally into the system, since, when once ingested in the body, it can not be eliminated and will produce radon gas within the human system for an entire life time and for hundreds of years thereafter. Experience of radium-dial watch workers in 1919, who became contaminated by touching paint brushes containing radium salts to their lips, proved the dangers of radium contamination. Uranium ores contain minute amounts of radium, so that protection of the uranium miner lies in simply washing the hands to avoid oral contamination in eating and drinking or smoking. The same rule holds true for the casual mine visitor. Care should always be taken to avoid drinking untested water in and around uranium mines.

The third element under discussion, radon, is a gas resulting directly from the decay of radium. Radon has an atomic weight of 222 and is known as element 86. It is colorless, and as a gas, is several times heavier than air, and, as an element exceeds the atomic weight of gold. It is odorless, and when present in the air of uranium mine workings can be detected only with radiation instruments. When present in heavy concentration it will completely permeate the working parts of an unsealed Geiger Counter, making such instrument temporarily valueless for accurate radiation readings.

Visitors at the Free Enterprise Mine breathe a limited and controlled amount of radon gas within each one-hour visiting period. While underground such person breathes out about as much radon as he breathes in. However, while radon is contained in the lungs, part of the radon gas is breaking down or decaying first to polonium (Po^{218}), element 84, with a half-life of 3.05 minutes; this element in turn decays to lead (Pb^{214}) with a half-life of 26.8 minutes; the next transmutation element appears as bismuth (Bi^{214}) with a half-life of 19.7 minutes; next, polonium (Po^{214}) with a half-life of 1.5×10^4 seconds; then thallium (Tl^{210}) with a half-life of 1.32 minutes. The half-life period is the time required for half of the atoms of a radioactive element present at the beginning to become disintegrated. The total disintegrated. The total half-life of the above five elements is less than 51 minutes, so that minute amounts of these elements are formed as solids, very short half-life isotopes, which are absorbed into the blood stream. These radioisotopes have been called miniature X-ray machines. Most certainly these must reach the billions of individual cells of the body, and likewise must reach the individual glands of the body. Whether one particular natural radioisotope, or a combination of several stimulate the glands to more active or normal production of body hormones is not known, but the fact remains that technicians have advised that such internal radiation does appear to stimulate certain glands to possible greater activity, and most certainly mine visitors who claim benefits, relief, or cures, are those afflicted with diseases which are said to be glandular connected.

From the foregoing it becomes evident that uranium, radium, and radon are separate and distinct elements, with quite different physical and chemical characteristics. Most certainly any criticism emanating from any source should differentiate in discussion of them. All uranium ore in its decay process produces radium, and radium in turn produces radon. But not all uranium mines produce, replace and maintain a constant supply of radon from day to day, as apparently do the Free Enterprise and the Uranium Mountain Mine.

Incidentally, a simple test for radon gas in a mine is for a mine visitor, on return to the surface, to breathe nasally on a Geiger Counter tube. Radioactivity acquired is reflected in the breath, with sufficient radiation to raise the dial needle of the Geiger Counter within the 2.0 Mr/Hr (milliroentgens per hour) scale. The visitor's hair and clothing, especially wool clothing, will be saturated and likewise show radioactivity for hours after mine visiting.

Recently, one man owning a uranium mine in Colorado is reported as providing, in all honesty and sincerity of purpose, accommodations for afflicted people. After receiving 300 visitors without a single claim of benefit he closed the mine. Quite evidently attendant and required physical, chemical, and geological factors in the mine were not present such as to induce the accumulation of radon.

Requirements for an effective "radon mine" are many. First, and of primary importance, ores within the mine must contain uranium, the parent element necessary to produce daughter products such as radon. Such uranium ore may be exposed in the developed mine workings, or may not be physically exposed but be known to be present at considerable depth, undisclosed and undeveloped. The second requirement for the presence of radon gas in the Boulder, Montana area is a favorable geological structure. This total area is intruded by a granitic mass known as monzonite, a phase of the granite. At the Free Enterprise Mine this monzonite is fractured and cross-fractured both as to the lode system and the country rock itself containing many cracks and openings, the whole near--surface granitic mass being an oxidized, decayed, and weathered mantle overlying the deeper, fresher, less altered granite. Uranium is contained not only in the veins and lode systems traversing the monzonite, but is likewise contained in small amounts in the attending country rock, the monzonite and numerous aplite dikes and masses. The "country rock" is presumed to contain 1 or 2 grams of uranium in every ton. These millions of tons of nearly barren rock all make a contribution to the radon gas total which finally reaches the underground workings of the Free Enterprise Mine. Since radon is a very heavy gas it is assumed that such gas is forced upward daily through cracks and fissures, by reason of increased pressure from beneath. Obviously, the whole process has taken billions of years to establish a constant or nearly constant production and replacement of radon gas. Every phase of decay in the uranium transmutation series is, at the present time, proceeding and continuing in the creation of radioisotopes of nature.

A national magazine in July, 1953, remarked that Montana mine owners hit pay dirt by charging crippled and credulous persons for sitting in worked out uranium mines and absorbing "radiation". The article further remarked that Nevada casino operators, figuring they were missing a bet, are now doing a goldrush business in Arizona, charging visitors "$3.50 for 30 minutes in an air-cooled shack." It further stated that Geiger Counters indicate the radiation is comparable only to that from an old radium dial watch. One technical society is quoted as calling it "An unfortunate hoax."

The author sees no possible benefit from such "uranium cures," which were promoted by the scores over many western states, places established under artificial conditions and aping the original uranium-radon discovery at the Free

Enterprise Mine. The unfair criticism is to group the reported uranium frauds with the now nationally and internationally known Montana property.

It is tragic that the head of a responsible institution, as quoted by a Los Angeles paper should be lured into making the following statement:

"Thousands of people have paid admission fees to abandoned western mines, lured by testimonials alleging cures of arthritis and related conditions by exposure to uranium."

Again this departmental head has missed the important point: Uranium, per se, apparently has nothing whatsoever to do with the benefits claimed by these hundreds of afflicted people. Radon only, the sixth decay element, of uranium, and the transmutation elements following radon gas in the uranium decay series appear to be the only elements which could be responsible for physical benefits and results reported.

As to relief afforded arthritics or others by reason of visiting the Free Enterprise Mine being entirely the result of "psychosomatic" influences, the manager of the Free Enterprise Mine office which kept daily documented records reported that about 80% of mine visitors claim some degree of relief or benefit, when the visiting process is given a fair trial. Most certainly a small percentage of any group of people may be listed as subject to psychosomatic influences. It appears doubtful, however, that a large percentage of visitors for such reason would claim benefits, when, in fact, many of the sufferers are initially skeptics at time of first visiting.

Normal or expected remissions in afflictions,

of course, are often present, coinciding perhaps with the time of mine visiting.

If all favorable reactions claimed by mine visitors be of a "subjective" nature, the benefits so claimed would be, those derived from brain and sense organs, and not from external stimuli. The former have been classified as fanciful, illusory reactions of persons excessively and moodily introspective. In some cases this might be true, but, on the other hand, certain scientific findings, mass evidence and testimonials reported, do imply that internal radiation, however nominal, induces catalytically certain effects relating to body glandular changes and physical improvement.

The charge that benefits resulting for arthritics and others were due entirely to the change to Montana's Rocky Mountain altitude of approximately 5,500 feet (mine elevation) is not tenable, since the first 800 to 1,000 visitors entering the Free Enterprise Mine came largely from local Montana towns or areas of nearly the same or higher altitude.

Additional years of observation and research have followed first publication of this book. Such research has resulted in a mass of additional evidence indicating that physical benefits claimed by thousands of mine visitors are not of psychosomatic origin, rather, the result of glandular stimulation and body chemistry changes. Evidence, thus acquired, is now in preponderant amount. The subject of radon therapy as provided by nature, with its attendant powerful factor of ionization, is now one, not of conjecture, rather one meriting further careful study and research.

CHAPTER 10

NATURAL AND ARTIFICIAL RADIOISOTOPES

Natural radioisotopes found in the mine air of the Free Enterprise underground workings include radon, Ra^{222}, a gas, the only radioisotope of the uranium series that is a gas. Those transmutation products which follow radon are solids, with alpha, beta, or gamma radiation. The next five decay elements immediately following radon are polonium, Po^{218}, lead, Pb^{214}, bismuth, Bi^{214}, a second polonium, Po^{214} and thallium, Tl^{210}. These five have a total half-life of less than 51 minutes. Older generic symbols designated them as RaA, RaB, RaC, RaC' and RaC", respectively. Laboratory radiation determinations for each of these elements still use these symbols, in connection with trapping and analysis of these breakdown elements from radon gas. The next elements of series decay, RaD, RaE, RaF, and RaG represent, respectively, lead, Pb^{210}, bismuth, Bi^{210}, a third polonium, Po^{210}, and final lead, Pb^{206}. The last lead is nonradioactive and stable. The half-lives of the last bismuth and polonium are measured in days, while lead, Pb^{210}, has the longest half-life, 22.2 years.

Besides the natural radioactive isotopes found in uranium, thorium, actinium, or other transmutation series, other mildly radioactive isotopes, or radioelements, have been found and identified, such as potassium 40, selenium 79, rubidium 87, indium 113, tin 124, tellurium 130, neodymium 150, samarium 152, lutetium 176, rhenium 187, lead 202, and bismuth 209. The half-life of most of these radioelements extends into the thousands or millions of years. Potassium 40 (K^{40}) is normally present in the human body to extent of about 0.012% of the total potassium contained in the human body. It is acquired in food we eat and from liquids, like milk, we drink.

In addition to radiation exposure acquired in body chemistry at intervals in a person's lifetime, everyone is subjected to cosmic ray bombardment. Additional dosages of radiation are acquired by many people through the chest X-ray, fluorograph of the chest, pelvimetry, abdominal fluoroscopy, and dental film. These dosages are reported not alarming, but technicians suggest extreme care in all instances and the keeping of individual records in matters relating to cumulative exposures.

Another type of exposure to which the whole world hopes we may not be subjected is that resulting from the A-bomb and the H-bomb, which two decades ago existed as theories only, but which today are threatening realities. Here in Montana the surface of many square miles covering many counties was some years ago made radioactive for days by reason of radioisotopes being carried northward on air currents following the explosion of an atomic bomb in Nevada. These radioisotopes resulting from government proving ground tests are reported of short half-life duration, harmless, but they did add a modicum of radiation to the land and water areas for several days following the bomb explosion.

Brief mention may be made of fissionable material and products resulting following the bomb burst. In the production of plutonium, an atomic bomb material, uranium, element 92, U^{238}, is transformed to U^{239}, which in turn becomes neptunium, a new element 93, Np^{239}, the atomic number being increased by 1. Neptunium being unstable, having a half-life of only 2.3 days, results in beta decay until this temporarily existing element becomes plutonium, atomic element 94, Pu^{239}. This important element, with alpha radiation, has a relatively long half-life of 24,000 years. U^{235} is likewise a fissionable material as is Pu^{239}. The use of these fissionable products in the bomb is dependent on critical mass and fusion, which factors, of course, represent secret information relating to atomic bomb explosions.

It is reported that nuclear fissionable elements have produced 200 different radioisotopes representing perhaps 34 elements. Many have short half-lives measured in minutes, hours, or a few days. There are others with relatively long half-lives. Resulting radioisotopes reported identified include iodine 131, 133, 134; barium 139,

140, 141; tellurium 127, 132; molybdenum 99; strontium 89, 91, 92; cerium 141, 143, 144; antimony 125; cadmium 115; and yttrium 91, 94.

In medicine, tracer radioisotopes have recently opened a new field in diagnostic investigation. The iodine radioisotope, I^{131}, is commonly known by its established use relating to the activity of the thyroid gland. I^{131}, with a half-life of about 8 days, when administered, localizes in the thyroid and thereafter the uptake of I^{131} may be measured externally with radiation instruments. I^{131} is perhaps the best known radioisotope used today.

Phosphorus, P^{32}, with a half-life of 14.3 days and cobalt, Co^{60}, with a half-life of 5.26 years are in standard use. The latter radioisotope is reported as a substitute for radium in intracavitary and interstitial applications, and its high radioactivity is reported comparable to an X-ray machine. The handling of Co^{60} requires the same care as the handling of radium. The daughter products from Co^{60} are solids, not gaseous as in the case of radium. Co^{58}, with a lesser half-life of about 72 days is used in tracer studies.

Strontium, Sr^{90}, has its applications in eye treatments. Radioactive colloidal gold solutions have been employed in certain diseases. There are a number of gold radioisotopes, Au^{198} being the most widely used. Radioactive calcium, Ca^{45}, is employed, but it is reported its use is not repeated for two years. It is also stated that when P^{32} and Ca^{45} are employed, a person should not at the same time be receiving radiation from other sources. Obviously the use of the artificial radioisotopes is the function of specialists in that field.

In March, 1953, newspapers reported that the Argonne Research Hospital of Chicago had opened, equipped to offer every known type of radiation thought useful for treatment for mankind. One dispatch described the new facilities, in part, as follows:

The new hospital has thick walls and many rooms are heavily shielded. Two of its eight floors are underground as a precaution against radiation hazards. It has so much intricate equipment, so many laboratories and treatment rooms that bed space for patients will be limited to two floors.

For treatment the hospital will use atomic particles or rays from these sources:

Speeding protons from the university's 450,000,000-volt synchrocyclotron, X-rays and electrons from a 2,000,000-volt Van de Graaff generator, electron "bullets" from a 50,000,000-volt linear accelerator, a cobalt "bomb" that can be rotated around the patient, radioisotopes from atomic piles, a 250,000-volt X-ray machine and smaller X-ray equipment.

Radioisotopes which will be used in treatment and tracer studies include carbon 14, tritium (radiohydrogen), radon gas, radiogold, radioison, radiophosphorous, radiochromium, and radioiodine.

Bed space has been provided for 56 patients on the third and fourth floors. Individual rooms are separated by eight-inch concrete walls. Floors are plastic-covered for ease in removing radiation contamination.

Persons entering rooms where radioactive contamination may be spread will be required to wear gowns, plastic overshoes and rubber gloves, all washable.

The radioisotopes will be stored in two-foot-long lead containers which fit inside stainless steel tubes extending eight feet below the basement floor and encased in solid concrete.

The hospital's facilities will be available to the Argonne national laboratory and the 32 Midwest universities and other research institutions which are participating members of the laboratory.

It might be particularly noted that "radon gas" is included in the list of radioisotopes to be used in treatment and tracer studies. Radon gas may be produced in the laboratory in limited amounts, but since a generous supply of radon gas exists in the underground workings of the Montana Free Enterprise Mine, it is suggested that such supply, in part, might be utilized in scientific studies and research, if mine operators and qualified scientific research personnel could meet in accord for cooperative and useful purposes.

To date the medical profession has utilized three sources of radiation. A professor of roentgenology has stated that its radiotherapeutic armamentarium today consists of first, X-rays; second, the radiations of radium; and, third, radiation of the artificially produced radioisotopes.

Now there may be added to the three major sources of radiation, a fourth, namely, natural radioisotopes, such as are found in the mine air of the Free Enterprise Mine. Based on studies undertaken over the past years, medical doctor reports, clinical evidence, and statements made by thousands of afflicted people, testifying to relief or cures for many afflictions, over the past

12 years, it appears reasonable to believe that this fourth source of radioactivity, supplied by nature, should now prove quite as acceptable as any of the other three methods, heretofore and at present employed. Most certainly as much time should now be devoted to the study of natural radioisotopes as to artificial ones.

CHAPTER 11

FREE ENTERPRISE RADIATION AND TOLERANCE

Effects of radiation on the human body are based on amount of radiation, frequency, its kind and intensity. X-ray technicians control the application of X-ray therapy, and doctors limit time period carefully in using radium externally. Tracer radioisotopes represent a third phase of radiation with which may be included artificial radioisotopes now used extensively in this new field.

The study of natural radioisotopes such as are found in the underground mine workings of the Free Enterprise Uranium Mine, and which represents a fourth phase of radiation, is becoming a specialized work in the radiation field.

The Free Enterprise operators were informed in 1952, through one department that a Committee on X-ray and Radium Protection had adopted a maximum allowable concentration, for an 8-hour working day, of 10 micromicrocuries of radon per liter of air. Studies were made by another agency indicating that it would be impossible for the mines to operate at this level, and it was therefore suggested that a tentative standard be established of 100 micromicrocuries of RaA (polonium 218) and RaC' (polonium 214) per liter of air. It was pointed out that uranium mines should have no difficulty through adequate ventilation of reducing the concentration of these radon decay elements to the level of 100 micromicrocuries.

Tests at the Free Enterprise Uranium Mine indicated a level of 1,450 micromicrocuries at the 85-foot level. The RaA and RaC' factors were determined as 720 micromicrocuries on the 85-foot level, the only level on which mine visiting is permitted. No exposures are permitted on the 150-foot level where level of radon concentration is reported as ... 10,220 micromicrocuries.

The radiation tolerance for mine workers has been tentatively suggested as 100 micromicrocuries as a matter of practical mine operation. It should be noted that such radiation would apply to the 8-hour shift worker, week after week, perhaps into years of continuous employment.

In visiting the Free Enterprise Mine for arthritis, sinusitis, asthma, or other affliction, you are interested in securing maximum relief. You are interested in knowing how many hours of radon exposure are recommended from any source. Advice in this regard comes to the operating company from doctors at Bockstein/Bad Gastein, Austria. The bulletin by Dr. Otto Henn regarding the radon adit there states as follows:

"Der radioaktive thermalstollen in Bockstein ist charakterisiert durch den Gohalt von 2.2-6.2x10⁻⁹ c Radium Emanation pro liter stollenluft. . ."

Translated freely, mathematically and chemically this means:

The radioactive thermal drift at Bockstein (Austria) is characterized by radiation factor of 2,200 to 6,200 micromicrocuries per liter of drift air.

Mine air and radon gases contained in the air of the 85-foot level have been trapped and tested likewise at the Free Enterprise Mine. Chemical and radiation test returns present a factor of about 1,450 micromicrocuries per liter of drift air. The average factor used at Bockstein is, as above noted, about 4,200 micromicrocuries.

Radiation exposure is measured and considered on the basis of *time and intensity.*

At Bockstein/Bad Gastein, Austria, doctors report permitting exposure over a series of at least 21 hours based on foregoing noted radiation factors. On the basis of 4,200 mcc to the Free Enterprise 1,450 mcc, 21 hours at Bad Gastein adit or underground workings appears to be the equivalent of about 51 hours at the Free Enterprise Mine, calculating on basis of inverse ratio. On basis of radiation exposure of 2,200 micromicrocuries of Bockstein/Bad Gastein the minimum time exposure equivalent for the Free Enterprise would be 32 hours.

Those claiming most relief and benefits after visiting the Free Enterprise Mine have taken a series of 12 *to 32 one-hour visits* as a rule, *two visits each day* — one in the morning and one in the afternoon. Those visitors taking less hours' exposure usually write in and say, "I have received benefits but feel that I would have benefited much more had I visited longer."

Initially this company was primarily interested in making certain that radiation underground at the Free Enterprise was well within tolerance limitations; that has been a safe rule to follow, and, based on times of exposure for the casual mine visitors, the company management believes that at all times the company has operated within tolerance limitations. The company is guided at all times by technical advice it receives from all reliable sources, both in the United States and from medical doctors of Europe. Where no contraindications appear, it would seem that from 12 to 32 hours' exposure would, according to advices received from the doctors of Europe, be at all times well within tolerance limitations for the mine visitors.

Advice received from a scientific departmental source in Salt Lake City, Utah, has stated the radiation in the underground atmosphere of the Free Enterprise Mine at the 85-foot level is so mild that the visitor could stay underground 24 hours per day for 30 days without reaching radiation tolerance limitations.

It might appear from reports received that the amount of radiation contained in the atmosphere at the Free Enterprise Mine may for a steadily employed miner, year after year, be considered above tentatively accepted tolerance limitations, providing, of course, he worked without proper ventilating conditions. For the casual visitor, in matter of total radiation exposure of a few hours only, it appears an entirely safe procedure. There is obviously a vast difference of a maximum 32 hours for the visitor and a possible 3,200 hours for the mine worker.

With reference to the biological effects of low-level irradiation some surprising results have been attained through several sources of investigation. A group of research workers in the Bio-Physics Department of the University of Washington Medical School in Seattle conducted a study with white rats. This group exposed 50 male rats continuously over their life span (which is 3 to 5 years) to controlled Cobalt 60 radiation for 8 hours per day at temperatures of 5°C and 25°C. An equal number of non-irradiated rats were used as controls. Three effects were then measured in the two groups of animals — oxygen consumption, mortality and histology (tissue study). Results were very surprising — the irradiated animals lived over 20% longer than the non-irradiated controls! Anticipated tissue damage did not occur. Experiments through other sources demonstrated increased longevity in animals exposed to low levels of radiation. These findings suggested "a general stimulating effect of an obscure mechanism of the radiation."

Additional research on the flour beetle concluded "the life span of a given number of flour beetles may be extended by several per cent by irradiation with gamma radiation."

While above experiments do not parallel the type or amount of radiation employed in the Free Enterprise Mine, it does appear that low-level radiation as employed through radon can be beneficial, and should not be compared to high-level radiation with its damaging aspects.

CHAPTER 12

ROMANCE, SEX LIFE: ARTHRITIS

Natural radioisotopes, radon and its transmuted elements, found in the air of the underground workings of the Free Enterprise Mine, represent, to some, a relatively new field of study. Most of the observation and research over the past decade has been done through operators of the mine and more recently, through Radon Research Foundation. Claims as to health benefits have been volunteered by thousands of mine visitors. Now doctors and technicians have supplied many reasons why these claims are made by the layman.

Advisors now state that if the boss pituitary gland, through nature's radon therapy is, through a modicum of internal radiation, stimulated in its production of ACTH, the entire endocrine gland system may be stimulated toward hormone production. It is concluded that the adrenal glands, the prostate, and likewise the reproductive glands may be benefited because the body's general well-being is effected.

Here it appears worth noting the statements of those who visit the mine and who volunteer particular information, vital to normal living.

For example, the case of the 33-year-old married man, a Montana resident, who visited the Free Enterprise in January, 1952. Summarizing, in substance, his own statement:

I visited the Free Enterprise Mine in January, 1952, spending one hour per day for two successive days in the underground mine workings. I had been afflicted with arthritis for several years, pain and swelling extending to numerous body joints. My neck stiffened from the beginning of calcification, until I could not turn my head without moving my entire body. By reason of the affliction I had been totally incapacitated so far as living a normal life in reference to marital sexual relations. Up to the time of my first visit to the Free Enterprise I had been deprived of normal functions for a one-year period.

Three weeks after visits I reported to one of the officers of the Elkhorn Mining Company, operating the Free Enterprise Mine, that I had received vast relief from the symptoms of arthritis, that calcification of my neck had broken loose, and moreover I was again normal in respect to functions of married life. Doctors explained that if the radiation factors present at the mine, due to radon and other decay elements, did stimulate the master pituitary gland to greater activity in the production of ACTH, that in that event such stimulation might tend to bring other body glands into greater activity or normalcy, including the genitals. I know that the mine helped me in these respects and I reported such startling benefits to those operating the Free Enterprise Mine for the sake of science, if a scientific explanation could be found.

Another arthritic, a young man in his twenties, visited the mine a few weeks later. He suffered pain, swelling in the body joints, and walked on the side of his foot. A few weeks later he reported freedom from pain, decreased swelling, and a straightening of the foot to the extent that he walked again in a normal manner. The young man was again permanently employed in gainful occupation. His wife had divorced him, presumably because of the arthritic affliction.

Inquiry into cases of arthritis afflicting youthful persons usually discovers a background of physical, mental or emotional shock, or a combination of them. Take the case of a 23-year-old boy suffering from severe arthritis symptoms, much pain, swelling, and attendant crippling condition. He told his story.

He had been engaged to be married, had providently purchased a new house, furnished it completely, and then his fiancee changed her mind, and married another. Deeply in love, the emotional shock to the boy was so great that life seemed meaningless. Arthritis developed, no body function seemed normal, and progress of the symptoms was rapid. Most certainly the arthritic condition had its mental and emotional beginnings, but the objective results were indeed real.

Another human interest story and tragedy revolved around a beautiful young woman who arrived at the Free Enterprise afflicted, according to her doctor's diagnosis, with rheumatoid arthritis. She was 32 years old, and had been afflicted since the age of 22. The author asked her what may have started her arthritis — was it due to mental, emotional or physical shock? Frankly, she responded, "All three" and then told her story:

"At the age of 22 I was very much in love, planning marriage. Without notice, my fiance married another girl. Disappointed and chagrined, I was "caught on the rebound," and became engaged to another man, a salesman, somewhat older than myself. I was not in love with him. One evening I was driving his automobile and crashed, receiving physical shock and injuries. This finger on my right hand is crooked because of the auto wreck the other three are gnarled by reason of arthritis.

"A few weeks thereafter, before I had time to recover from the physical shock, I married the salesman. He turned out to be unkind cruel, and intoxicated much of the time. My health suffered, symptoms of arthritis developed, and progress of the disease was rapid. Answering your questions, my present condition may be ascribed to three factors: physical, emotional and mental shock."

One doctor has stated that all arthritis has its psychosomatic origin. Since life is mental and physical, with mind and body closely allied as a functional unit, physical troubles will evidently continue to recur as in the individual cases noted.

Visitors at the Free Enterprise Mine have claimed relief and benefits for quite a number of ailments, including arthritis, bursitis, sinusitis, asthma, multiple-sclerosis, eczema and other skin afflictions. They claim benefit for a wide scope of afflictions where the diseases appear related to glandular functioning. After one woman claimed freedom from eczema following a few visits to the mine, a Beverly Hills specialist was asked if skin afflictions were glandular connected.

"Most of them are," he replied. Another woman visitor claimed she lost 16 pounds within a few weeks after mine visiting, claiming that in her case excess weight had been due to glandular trouble, and that the modicum of internal radiation acquired at the mine stimulated her glands to greater or more normal activity. Others reported improvement in sight and hearing. Among the many claims by mine visitors to date could there be added increased potency and fertility in individual cases resulting from improved general well-being?

Four local Montana cases have reported pregnancies following radon mine exposure. The first couple had been married 9 years without offspring. After visits to the Free Enterprise radon discovery a child was born within the year. The father, only, has visited the mine.

The second similar case represented a couple with only one child, with a negative reproductive period of six years following. After employment of the father in a mine that contained radon, the couple reported a second child born within 13 months.

The third case concerned a childless marriage of 17 years duration. The father was a miner at a property having a fair degree of radon present in the mine workings. In the eighteenth year of marriage he became a father.

On June 2nd, 1954, a young Montana man

reported to the author the following story:

"In April, of 1948, my wife lost twin boys through premature birth. From that date into June of 1950 she had been unable to again attain pregnancy. Eventually, a third child, a boy, was born in March, 1951.

"In the early part of April 1952 my wife's doctor ordered an operation. Conference at that time was had between my wife, myself, and her doctor who advised that the operation could be performed first eliminating the possibility of more children, or second, could be performed, leaving the possibility of further pregnancies. I favored the first alternative for my wife's protection. Although we had three, my wife desired more children.

"Up to April 1952 my wife's condition indicated that another pregnancy was not possible and undesirable from a health standpoint, with an operation recommended by her doctor.

"Finally in the latter part of April of 1952, I suggested to my wife that if she could not become pregnant by June of 1952, that she should consent to the operation eliminating the possibility of having more children. She agreed to this suggestion.

"In May of 1952 she visited a radon mine in the Basin-Boulder area, and believe it or not, she became pregnant the next month, in June. It was a boy. Now our family consists of two boys and two girls. Incidentally, an operation was avoided.

"Whether or not the visit to the uranium-radon mine induced normalcy for my wife, I do not know; I am simply reporting the facts in our individual case. I have, however, heard of similar cases."

These four cases may, of course, be wholly coincidental to environmental conditions, but are reported for record purposes.

One important observation should be mentioned. Quite a number of men afflicted with prostate gland difficulties have visited the Free Enterprise Mine. A high percentage of these cases, after a series of mine visiting, have reported decreased swelling of that gland with consequent relief from their problems. Some, stating they were due for hospitalization, claim they avoided surgery.

Many treatises testify first to the benefits of X-ray therapy, and second to the dangers unless administered by qualified radiologists. The benefits are well known to everyone. It is also common knowledge that temporary or permanent sterilization may occur by overexposure in pelvic regions of the body. Extreme radiation from the atomic bomb burst or other radiation sources have their many adverse effects on life and health. Quite apparently radiation is good medicine or bad medicine, dependent on dosage in the sense of degree or intensity of application or use.

The element radium has had, for a long time, its therapy uses. More recently artificial radioisotopes handled by those qualified in their use, are being employed in increasing quantities.

As mentioned, X-ray, radium, and artificial radioisotopes, three phases of radiation, are in common usage today. Certainly, the fourth phase of radiation, natural radioisotopes, is worthy of exhaustive study for the relief of afflicted people and the happiness of mankind. Based on voluntary statements of thousands of mine visitors for over a decade, their independent statements paralleling and conforming to a specific pattern, it now appears that such statements, together with other findings, are worthy of consideration, merit scientific investigation, and now warrant a period of clinical evaluation studies.

.

CHAPTER 13

VISITORS' FACILITIES AT BOULDER, MONTANA

First, the mine visitor is interested in finding a real uranium mine, one containing radon gas, to which element benefits have been attributed. Second, one is likewise concerned with good housing facilities — reasonably priced. Many afflicted people have spent thousands of dollars with their doctors on medications and treatments for arthritis or other ailments over the years. Most must, of necessity, economize. While arthritis strikes the rich as well as the poor it must be recognized that the greater majority of afflicted people have, through medical expenses over the years, depleted their capital.

From a variety of facilities, varying in price, accommodations may be chosen. All motels and restaurants in Boulder, Montana and vicinity, have cooperated with the Free Enterprise Health Mine in maintaining reasonable charges for room and meals. Within the town of Boulder are: the Linn Motel — with easy access to Interstate 15 and a short drive or walk to the heart of town; the O-Z Motel — situated on Main Street, in the middle of town, is near restaurants and grocers; and the Castoria Inn, just blocks off the main street, is in a neighborhood setting. Three miles south of town, the Boulder Hot Springs (formerly called the Diamond 'S' Ranchhotel) is undergoing extensive renovation at this time. They currently have a number of bed and breakfast rooms available to the public — to include soaking pools. In addition, indoor soaking mineral pools and steam rooms are open to the general public on a limited schedule during the week.

The housing facilities noted above are usually sufficient to meet public requirements except at intervals during June, July, August and September, when summer travel and Mine visiting is at its peak. Make reservations, well in advance, at the motel of your choice.

Other businesses in town include two grocery stores, several restaurants, a drug store, gas station, post office, library, bank, bowling alley, coin laundries, car wash, senior center, medical clinic, movie video rentals, auto repair, weekly newspaper and churches in a variety of denominations.

Last, but not least, the out-of-state Free Enterprise Health Mine visitor is assured of Montana hospitality known the world over. There are no strangers in Montana. In the pioneer days people used to say that Montanans didn't die, they just got killed occasionally. It is not that rough anymore. Everyone is friendly and will speak to you, whether you have met before or not — not a matter of being forward, simply a gesture of genuine friendly spirit. In Montana, the measure of one's wealth is not a measure of one's worth or importance in the community. A day laborer may see his Governor. In Boulder, Montana, local people work remarkably well together, in complete harmony. Visitors will immediately recognize their spirit of hospitality, and become a part of it.

Always be prepared for an off-season snow storm! The Free Enterprise Health Mine, at this writing, is open seasonally from March 1 through November 30. No reservations are needed for visiting the Mine. Hours of operation are 8 A.M. to 6 P.M., seven days a week. The recommended schedule of visits is 3 1-hour visits per day, with a minimum of 2 hours out of the Mine between visits — resting as much as possible. The visitor will need to allow at least a 10 day stay in Boulder for the recommended 30-visit schedule. This schedule is especially important for the first time visitor to allow the benefits of radon to take its course. Many Mine visitors will find that taking half the maximum time, twice per year (Spring and Autumn) proves to be just as beneficial, if not more so.

Upon initially entering the Mine lobby, each individual registers with the receptionist and receives a ticket for visits purchased in advance. This ticket has the dual purpose of being a receipt and a punch card that allows visitors to keep track of their time in the Mine. The recommended stay is based on mandates set by the State Department of Health and Environmental Science in Helena, Montana in cooperation with the Free Enterprise Health Mine. The Free Enterprise Health Mine complies with all standards set by this and other state agencies.

After descending 85 feet in the original Otis elevator, the Mine visitor enters the 400 feet of horizontal drift, excavated in 1950-52. Cushioned seats are situated under lights with heat lamps at frequent intervals. An auxiliary generator is the backup in the event of a temporary power outage and emergency lighting takes over immediately.

A recent addition is the 'Inhalatorium' (radon room) located on the main floor, to the rear of the office lobby. The radon gas is pumped from the mine workings at the 105 foot level, filtered, circulated in the room, then returned to a lower level of the mine creating conditions for radon therapy as near as possible to that of the mine itself. A warmer, living room setting also adds to the comfort of this room for those who are uncomfortable descending underground.

Restrooms, telephone (collect or credit card only) and water facilities are located on the main floor. Two handicapped accessible showers have been added for the use of those staying in the RV campground on the premises. (See a current brochure for details.)

The visitor will need to make their own travel arrangements to Boulder, Montana. The commercial airlines service Butte and Helena, Montana. Check with the motel proprietor for transportation to and from the airport and/or bus depot. Intermountain Trailways currently stops at the Town Pump in Boulder. The Free Enterprise shuttle van is available for those visitors who arrive by bus or airplane and who have not rented a car or do not have transportation. The current schedule — subject to change — is set for three round trips per day to and from the local motels in Boulder and a nominal fee is charged on a per-trip basis, payable at the end of your stay.

A current brochure with testimonials (complete with names, addresses and telephone numbers) and updated information will always be available by calling (406) 225-3383 or writing to:

FREE ENTERPRISE HEALTH MINE
P.O. Box 67
Boulder MT 59632

Request brochures for yourself or send to the above address a list of names and addresses and we will mail for you.

CHAPTER 14

THE NEED FOR SCIENTIFIC RESEARCH

Proper approach and study of the Free Enterprise Mine operation in its use relative to afflicted people embraces many sciences, including geology, mineralogy, chemistry, physics, radiology, ionization, and medicine. No one person could live long enough to become conversant with all the ramifications of science pertinent to proper consideration of the problems involved.

The pressing need at the present time is a scientific study of radiation yielded by natural radioisotopes such as are found in the air of the Free Enterprise mine workings. Much research has already been accomplished with reference to laboratory-created isotopes representing one particular direction of research only. These artificial radioisotopes are being used in medicine.

Assuming that radon therapy does stimulate the boss pituitary gland in its normal and continuous production of ACTH, one theorizes that radon mine exposure should afford relief to the leukemia victim on a more permanent basis. This theory and possibility was presented to three medical doctors in 1958 and all agreed that this observation could be highly important.

No leukemia case has ever visited the Free Enterprise Mine and none will be knowingly permitted without a medical doctor's order. It does seem worth a try, in view of the foregoing scientific reasons, especially in an acute leukemia case where life expectancy is limited to 4 or 5 months.

Many inquiries have been received asking if radon mine visiting can help multiple sclerosis, Parkinson's and similar afflictions. No claim is made for benefiting these ailments but many visitors so afflicted over the past years have claimed better locomotion and mobility of body members after a series of mine visits. Doctors explain that if glandular retardation accompany such cases, and that if the glands are stimulated toward normalcy in production of body hormones, benefits may accrue. Benefits are being claimed for a wide variety of afflictions and daily observations and experience point to relief for many more ailments than at first contemplated. In brief, most afflictions that are glandular connected appear to be benefited.

Some 15 years ago, cortisone, derived from animal gland or synthetic sources, was hailed as a miracle drug for relief of arthritis in its various forms. Now, however, with medical observations of the adverse side effects from administration of cortisone over a long time period, a majority of doctors do not administer the drug. Among the side effects are edema, ulcers, and hirsutism. Many visitors state they have had a portion of the stomach removed because of ulcers resulting from long term administration of cortisone.

Observations of hundreds of cases over the past 12 years of continuous radon mine operation indicate that those persons who have not taken cortisone, or its derivatives, respond more readily to radon exposure than those on cortisone. It is reported that administration of cortisone over a long time period hinders the body's normal defense mechanism by taking away the body's work responsibility of producing a better product, hydrocortisone. After all, the purpose of the Free Enterprise Mine visiting is to stimulate the endocrine gland system toward normal production of body hormones. No evidence, over the years, has come up to refute the theory that this is accomplished by internal radiation, ionization, catalytic action, or a combination of these environmental factors.

IONIZATION

One environmental factor that further research may prove highly important, attending Free Enterprise Mine therapy, is ionization. For a long time radon has been known to be a powerful ionizing influence.

In May of 1961 two medical doctors and an engineer arrived at the Free Enterprise Mine with equipment for determining the amount of ionization present in the underground mine

workings. Engineer's report to the mine operators stated that 500,000 ions of equal polarity
per cubic centimeter were contained in the underground atmosphere at the 85-foot level; that
50,000 ions per c.c. were present in the Free
Enterprise waiting room at the surface; that
5,000 ions per c.c. existed at one location within
the town of Boulder, Montana.

That the Free Enterprise Mine and environs
represents a unique area is evident by comparing
ionization determinations in other parts of the
United States. For example, natural ion levels in
the City of Philadelphia are reported to range
from 832 to 874 per c.c.; while the average
number of ions at Richland, Washington, to-
talled 1420, including both negative and positive
ions.

In October of 1961 an International Conference on ionization of the Air was held at the
Franklin Institute, Philadelphia, and sponsored
by the American Institute of Medical Climatology. The author, a member, attended. Igho Hart
Kornbleuh, M.D., Director of the Institute, organized the voluminous proceedings which were
resolved in two volumes. Dr. Kornbleuh is the
one person in the United States who has conducted extensive research on the subject of ionization and who may be considered a pioneer
and expert on determinations to date. Medical
doctors and scientists from Denmark, Sweden,
France, Russia and the United States attended
the conference and presented technical papers
on many subjects and phases relating to ionization.

The first paper, entitled "An Attempt to
Define 'Ionization of the Air'", was presented

by C.W. Hansell, R.C.A. Laboratories, of Princeton, New Jersey. Excerpts from this paper follow:

"Ion" is defined in Webster's dictionary as an electrical charged atom, or groups of atoms, in a liquid solution, or in gas.

Some difficulties with adopting Webster's definition of "ion" are that, first, it sets no limit on the number of atoms in the charged group of atoms. It would permit us to refer to the whole earth as an "ion" since it is a negatively charged collection of atoms in an atmosphere of gas....

Passing through the atmosphere are a relatively few particles, and damped electro magnetic wave trains, having velocities far greater than the velocities of the molecules due to temperature. Some of these are cosmic rays arriving from outer space and the secondary particles and radiations produced by them in the earth's atmosphere. Others are produced by nuclear disintegrations of radioactive atoms present in the earth and air of which the radioactive gas radon deserves particular note....

The human race was developed in ionized air, through a long process of creation and evolution. The published reports of air ionization biological research, and papers to be presented here, suggest that nature utilized the ions in developing our biological processes, and the marvelous system of negative feedback controls by which our metabolic rates are regulated. Russian experiments with animals, put in air deprived of ions as completely as possible, indicated a probability that we can not live without them.

There is another factor at work. Solid and liquid surfaces are very effective in removing the charges from ions which strike them, to convert them back into neutral molecules. On that account any increase in the ratio surface to volume of air tends to reduce ion densities. Indoor air almost always contains fewer ions than outdoor air. When rooms are occupied by numbers of persons their breathing strips the respired air of ions. At the same time the humidity is increased, which makes condensation nuclei more effective deionizers, while remaining ions become more heavily clustered. Cigarette smoke can add greatly to the densities of the condensation nuclei. Mechanical air circulating systems with their blowers, coolers, washers, filters, ducts, grilles, etc., can cause large losses of air ions. This provides a real basis for the complaint of many sensitive persons that "perfectly" conditioned air still is depressing them.

Because negative ions with their molecular clusters are smaller and more mobile than positive ions with their larger molecular clusters they tend to be removed at a greater rate at surfaces. Consequently surfaces exposed to air in which ionization is taking place almost always assume a negative charge, leaving the air positively charged. The largest example of this is the surface of the earth which is negatively charged while the atmosphere is positively charged. The charges are great enough to create a difference of potential, between the upper atmosphere and the earth, of about 300,000 volts. This is the potential required to bring the collection of positive and negative ions by the earth into balance. The same kind of phenomenon makes the air in rooms tend to become abnormally positively charged, or positively ionized.

There is still another important factor in our environment which should be taken into account. Our distant ancestors, in the course of our evolutionary development, lived, literally, with their feet on the ground. Their bodies were kept at ground potential. In contrast, we are electrically insulated from the ground much of the time. Often our bodies are at potentials far different from ground and our surroundings. These potentials can have large effects upon the ratio of positive to negative ions we absorb from the air and upon their total number. An outstanding example is provided in warm dry air, in winter time, when we walk about on clean new wool carpets in rubber or leather soled shoes. Our bodies then may become negatively charged to potentials of tens of thousands of volts. We can sometimes draw long sparks from grounded objects. Under these circumstances negative ions are strongly repelled by our negatively charged bodies but positive ions are strongly attracted. We then are likely to be physically, mentally and emotionally depressed and irritable. It would be better if we could put the rubber or leather on the floor and the wool on the feet.

Almost always, transient contacts between outer clothing and room or furniture surfaces result in transfer of charge and build up of body potentials of one polarity or the other so that absorption of ions from the air is affected. A few people respond so strongly that they are forced to notice that they cannot tolerate some types of materials. The time may come when clothing and room surface materials will be graded and labelled according to their positions in the triboelectric series. Some manufacturers of fibers and fabrics have already shown awareness of the problem. A wry thought for persons who suffer from such effects is that dirty surfaces seldom exhibit strong electrical charging effects.

For sixteen years since the advent of the atom bomb people have been deeply concerned about increasing pollution of the earth and atmosphere by radioactive materials. Renewal of nuclear bomb testing in the atmosphere by the U.S.S.R. has increased the fear of disastrous effects of consequent increases in genetic changes and mutations in living things, including people. At the same time there have been places, such as the radium baths of Europe and old uranium mines in the United States, where people have been paying for the privilege of exposing themselves to a relatively high level of radioactivity. The proprietors of these places, and many of their patrons, claim that the exposures provide important therapeutic benefits in a variety of disabilities and diseases with which the medical profession as a whole has not been very successful. Our government law enforcement agencies have been taking legal actions to close these places on the ground that therapeutic claims made for them are fraudulent. It is therefore interesting to note that there appeared in the Journal of the American Medical Association *a report that animals kept exposed to increased radioactivity were improved in general health and longevity, as compared with animals not so exposed. (Henry, J. A.M.A. Vol. 176, No. 8, P. 212-125, May 27, 1961.) Thus it appears that, although increased exposures to radioactivity produce genetic changes which may be generally harmful, moderate exposures, at the same time, may improve the general health and increase the average life span. I suspect that the health benefits due to exposures to reasonable increases in radioactivity are real, and that they result, at least in part, from the attendant increase in air ionization.*

We might now ask which of the abnormalities in respect to ionization of the air causes us the most discomfort and harm? Is it the unnatural preponderance of positive over negative ion densities which most needs correction, or is it the unnaturally low densities of ions of both polarities? The present manufacturers and marketers of negative air ionizers, judging by their actions and advertising, must have assumed that addition of only negative ions to room air was enough to make us healthy and happy. How much chance do you suppose nature gives them of being right? As for myself, for air conditioning, I would be prejudiced toward imitating nature's best more closely, by maintaining optimum densities of ions of both polarities. You who are doing air ionization biological medical research are expected to provide the answer.

The second paper, presented by Dr. Rudolph Nagy, Consulting Engineer for the Westinghouse Electric Corporation of Bloomfield, New Jersey, was entitled "Nature of Air Ions Generated by Different Methods." Dr. Nagy pointed out that "during the past sixty years many investigators have been using air ions in the treatment of respiratory and other types of diseases." He described five methods of ion generation: charge separation, thermionic emission, corona discharge, radioactive materials and ultraviolet radiation.

Professor A. A. Minkh, of the Academy of Medical Sciences, Russia, submitted a paper entitled, "The Effect of Ionized Air on Work Capacity and Vitamin Metabolism." Highlights of his observation: Through highly ionized air of negative polarity there is considerable improvement in the general tone, cheerfulness, energy, good sleep and appetite; muscular strength of the hands may increase by 16%; negatively ionized air has an antiscurvy effect in experiments with animals when deprived of Vitamin C; employed in medical practice in USSR, it increases the physical work capacity and improves the general tone of healthy people.

Some most remarkable observations were presented in a paper entitled, "The Effects of Negative Ions Produced by Fibers of Polyvinyl Chloride," by Christian Folger of Societe Rhoyvl, Meuse, France, being a summary of the work of four French researchers. Determinations:

That fabrics of pure Polyvinyl Chloride always develop considerable amounts of negative static electricity. Wool, silk and cotton do not. A definite analgesic effect is produced by wearing underwear, socks, and other clothing of this material. Bandages, sheets, blankets and coverlets for individual joints may be employed. The underlying cause of pain is not eliminated but in many cases marked relief or complete cessation of pain is achieved. The average decrease in pain of persons wearing polyvinyl chloride clothing is 16%. Improved peripheral circulation and sleep are frequently noted.

It was further noted that materials made from polyvinyl chloride fibre may be sterilized and used indefinitely. Likewise, because of its non-hygroscopic properties it is especially suitable for the manufacture of semi-elastic bandages and non-adhering dressings and compresses. Studies on this subject, originated in France, are being currently studied in Italy, Czechoslovakia and the Soviet Union.

Through Radon Research Foundation, the author has contacted several large manufacturing firms in the United States in an endeavor to find a source of this material, without results. It does appear the polyvinyl chloride fibre materials could now become a valuable adjunct and addition for use in hospitals and as added modern therapeutic armamentarium for our American medical doctors.

In a paper entitled, "The Clinical, Scientific and Technical Development of Electro-aerosology," Dr. Alfred P. Wehner, of Dallas, Texas, chose to discuss ions as attaching themselves to minute particles which are always copiously suspended in the air. These are then known as electro-aerosols, charged particles. He notes that in Germany electro-aerosol therapy has reached a considerably advanced state, that claims have been made for the following afflictions: bronchial asthma, bronchitis, pertussis, sinusitis, rhinitis, pulmonary emphysema, migraine allergies, arthritis, hypertension, painful and irregular menstruation, chronic eczema. Many other ailments were mentioned. The doctor admits that controversy has arisen regarding benefits claimed for such a wide range of afflictions, but here it is worthy of observation that similar claims for the same ailments have been made by Free Enterprise Mine visitors over the past 12 year period of continuous operation. Dr. Wehner stated the European scientists had expressed amazement that the United States had not yet utilized "this well-tested, effective, safe and economical therapy.

An important "Post Conference Comment By The General Chairman" of the 1961

Philadelphia Meeting was added to the published Ionization Proceedings by Dr. Clarence W. Hansell of the R. C. A. Laboratories, Princeton, New Jersey. Excerpts from this paper follow:

The amplifier system, or systems, in the living body comprise catalysts, called enzymes, which must be present and active to permit many of the biochemical body processes to take place. The enzymes are, in turn, activated or deactivated according to need by activators and deactivators called hormones. The hormones are, in general, manufactured in the ductless glands and stored in the glands and body tissues from which they are released at variable rates in response to variations in blood composition and flow, body temperature, messages from the brain and nervous system and other factors which signal a need for changes in metabolic rates. The complex of enzymes, hormones and associated organs is called the endocrine system.

Activation and deactivation of enzyme catalysts by the hormones provides a large degree of amplification in that relatively very small amounts of matter and energy, represented by the hormones, control relatively enormous amounts of matter and energy in the catalyzed biochemical reactions. The feedback in this system is provided through the effects of products of the catalyzed reaction upon the rate of release of hormones. The bias or operating level adjustment is controlled and varied by the effects of external environment and other stimuli, some acting directly upon the body and some acting through the brain and nervous system.

* * * * *

Among chronic abnormalities and diseases there is a large group in which the symptoms generally are attributed, or suspected to be due to "endocrine disturbances." From another point of view we can say that almost all injuries and diseases produce endocrine responses and disturbances, which in turn are responsible for many of the symptoms.

Author's note: Many persons, following physical injury, including auto accidents and surgery, with consequent shock to the endocrine gland system, eventually arrive at the Free Enterprise Mine to avoid arthritis and its initial symptoms.

* * * * *

Those living near sea level generally obtain relief when they go to higher elevations where ionization is greater and positive ions less predominant.

* * * * *

When either negative ions or cortisone are given to relieve disease symptoms they tend to compensate for hormone deficiencies which produced the symptoms. The glands or parts in the body, involved in providing the deficient hormone then no longer are required or stimulated to provide even the little hormone that they provided before. They are deceived into acting as if they were providing too much, rather than too little. They slowly adjust toward smaller output levels and output capability. In the case of cortisone, for example, its administration relieves the pituitary and adrenal glands of demand for their output. One result, reported in medical literature, is that patients who received too much cortisone for too long, upon autopsy, exhibited atrophy of the pituitary gland. Thus cortisone and negative ions, it would seem, should be used as sparingly as possible if required to provide temporary relief of symptoms and should be recognized as the opposite of a cure. After they have been given in too great quantity for too long they must be withdrawn slowly to avoid strong reactions.

* * * * *

The transient effects of ionized air may be utilized very beneficially in therapeutics under those circumstances where long term effects are not required. Such circumstances are found among badly burned and injured patients, including those who have just undergone surgery. Similarly, only transient responses may be required to initiate healing of ulcers and ulcerating wounds which are refusing to heal, and to give the body an advantage in resisting infections. These applications can be of importance to stomatologists and dentists, as well as surgeons and doctors.

* * * * *

The heating and air conditioning industry, although it has, as a whole, tended to ignore or reject the idea that its equipment exposes people to abnormal and depressing conditions of air ionization will surely be, forced to do something about it in the future. Our model suggests that the promotion of negative air ionizers for constant use in living spaces may have been to some degree an error. In so far as these devices are used to restore a more nearly natural level, and natural ratio of positive to negative ions they can be healthful and beneficial. In many circumstances they can restore a more nearly normal ratio of positive and negative ion densities and may even increase positive ion densities, as well as negative, by balancing out positive space charges which tend to drive the positive ions toward surfaces. In other circumstances they may subject persons to excessive absorptions of negative ions which, if too long continued, can lead to objectional body adjustment or addiction effects. It is the opinion of the author that heating and air conditioning engineers should not subject people to therapeutic doses of ions, but that they have a duty to restore natural, healthful, densities of ions of both polarities. They should correct the effects of their equipment upon ion densities which are deleterious to human health, comfort and efficiency.

* * * * *

As has heretofore been noted, radon is a powerful ionizing influence. Instrument tests made by medical doctors in cooperation with engineers at the 85-foot level of the Free Enterprise Mine have shown the presence of 500,000 ions per cubic centimeter. The positive and negative ion counts were about equal in number, which is a normal condition in nature.

Technical papers rendered by medical doctors and engineers at the International Ionization Conference, held at the Franklin Institute, Philadelphia, stressed the physical benefits to be derived from controlled exposure to negative ionization, produced by artificial means. Igho H. Kornbleuh, M.D., Director of the American Institute of Medical Climatology, has pioneered much of the research in the field of ionization,

especially in the application of negative ionization. In the author's opinion he is an expert in
this field and versed in both generalized and
specific knowledge relating to this science.

Several conference participants stressed and
held the opinion that for general application, as
contained in the remarks by Dr. C. W. Hansell,
they "would be prejudiced toward imitating
nature's best more closely, by maintaining
optimum densities of ions of both polarities."
As above noted, the underground radon laden
atmosphere of the Free Enterprise Mine does
maintain a very high density of ionization of
both positive and negative polarities, which
accounts for one important scientific reason for
the physical benefits and sense of well-being
experienced by a majority of visitors to the Free
Enterprise Mine.

Igho Hart Kornbleuh, M.D., Director of
Physical Therapy, University of Pennsylvania,
Graduate Hospital, Philadelphia, and Director of
the American Institute of Medical Climatology,
presented a paper entitled "The Future of
Artificial Ionization of the Air." Dr. Kornbleuh,
responsible for the success of the International
Conference on Ionization, has traveled
extensively in Europe, Russia, England, and
Africa, on research and appears to be the most
informed doctor and scientist on this subject.
He is conversant with the methods of ion
generation and in the application and proper use
of negative ionization. While successful in his
observations of its beneficial uses, especially with
reference to its pain relieving qualities, and
amelioration of severe burn cases, as has been
demonstrated in Philadelphia hospitals, he is

likewise cognizant of the importance of ionization as supplied by natural environmental sources.

It is not satisfying to make reference to a portion of an able paper, but here are quoted a number of his statements standing out as most salient.

We are all well aware that the physical characteristics and the biological effects of air ions are quite exciting but not fully understood, and appreciated.

* * * * *

Much effort and work are being wasted to undermine and discredit the value of natural resources and elements with which this planet is so richly endowed. Ionization of air is one of these. Being old-fashioned we have followed our beliefs only to be frequently embarrassed by the results of our own investigations. The embarrassment comes from the difficulty created by the lack of proper understanding for the interplay between the innate nature of the living organism and the physical environment. Modern, pseudo-scientific concepts put man on a pedestal to show his independence from his environment. In reality, his life, health and destiny depend entirely on the latter. Organic man is not and cannot be a stranger in an inorganic universe.

* * * * *

In this synthetic world man gradually surrenders his natural ways and becomes a greenhouse product well fertilized with vitamins, stimulated with coffee, alcohol and tobacco and carefully restrained with powerful tranquilizers. This alarming and deplorable tendency will not be dramatically changed solely by saturation of the air with products of the most sophisticated ion generators. Nevertheless, under the prevailing conditions, normalization by the means of inclusion of controlled artificial aeroionization is a step in the right direction.

The toil of a small but dedicated group of scientists in this country has established with certainty that ionized air has a definite place in our physical environment. The pioneering work of the Russian school has greatly contributed to our knowledge of the technical and biological aspects of aeroionization. Because of the geographical location, the climate of the Soviet Union does not require during the summer any corrective measures for sustenance of human comfort. This permitted the Russian investigators to concentrate their work on the biological, physiological and therapeutic effects of artificially generated ions. Their great advances in these fields must be acknowledged here.

* * * * *

The pain-relieving quality of negative friction electricity, in contrast to the ineffective positive one, will certainly mobilize our textile experts. It could be that our present day furnishings and garments, because of their static properties, will have to undergo drastic changes. Presentation of the therapeutic significance of charged aerosols directs our attention to a modified and, in this country, new method of ionization.

* * * * *

To assure fast progress a close cooperation of medicine and technology is needed. A better understanding of the aims and needs of medicine by the engineers and a grasp of the technical possibilities on the part of physicians will bring forth new diagnostic and therapeutic tools of great practical significance. The present interlingua understood by both groups has not yet attained the desired perfection. Centuries old medical traditions have never envisioned the possibility of such a development. The sooner the curtains and barriers dividing both branches of science are removed the swifter will be the progress. A close association on an equal footing and a common plateau is the desired goal. With good will, mutual understanding and respect this could be achieved in a not too distant future.

CHAPTER 15

CONCLUSION

These are the circumstances of the discovery of the Free Enterprise Mine — the first Montana commercial uranium ore producer, found June 18, 1949. Secondary decay elements of uranium, including radon, a gas, and its transmuted elements, it now appears, may be recognized as more important to the well-being of mankind than uranium itself.

The Free Enterprise Health Mine has now been open to the public since 1952. Thousands of afflicted people have passed through the Mine portals. Beneficial results have been real and not due to psychosomatic influences. Constructive criticism of the operation is always invited. Adverse criticism, without study or research, has been due to misinformation, or ignorance of the science of radon therapy and the important attending factor of ionization found in the mine's underground atmosphere.

In view of the scientific observations to date the following conclusions appear pertinent and evident:

1. Radon, in controlled amounts, with its transmuted elements produces a mild but effective internal radiation, reaching the blood stream, the body cells and the endocrine gland system, inducing through direct or catalytic action, the pituitary to produce ACTH, the adrenal cortex to produce hydrocortisone with evidence that other body glands are likewise stimulated toward normalcy.

2. Nature's foregoing treatment represents a scientific breakthrough — offering a remedy reaching a principle cause of arthritis and many kindred glandular connected afflictions.

3. Exposure to the powerful factor of ionization attending radon and its transmuted elements constitutes an analgesic and beneficial environment influence. This conclusion is concurred in by many doctors and scientists, both in Europe and the United States.

4. Many individual medical doctors have, through laboratory evidence, relative to the patients they have sent to the Free Enterprise Health Mine, determined the physical benefits to be derived from radon therapy and ionization. Many of these doctors have visited for their own afflictions and acknowledged benefits.

It now appears that more medical doctors, as well as those Foundations dedicated to aiding arthritics, should be enthusiastic in recommending these factors of nature that are now obviously benefiting suffering humanity. By reason of the irrefutable results to date we ask the question...

"Why remain in pain?"

ABOUT THE AUTHOR

Wade Vernon Lewis was born in Portland, Oregon to a pioneer family in 1893. His father, John M. Lewis, was Treasurer of Multnomah County for 30 years. His mother, Ella, was a MacPherson — one of Portland's earliest printers and newspaper families.

After elementary and high school, Lewis attended a year at Oregon State School of Mining and Agriculture (OSC). World War II took him to France and upon his return he completed his degree in Mining Engineering and Geology.

Lewis worked as a chemist and assayer. For twelve years, he was employed as a surveyor and mineral examiner for the U.S. Land Office. The lure of gold took him to Montana and the depression was in full swing. His desire to be a family man and his burning quest for finding the "mother lode" became an intense roller-coaster odyssey of several decades. The following years became a series of trials and tribulations. He also suffered a short bout of polio. The satisfaction came with the mining of silver and pitchblende from the Free Enterprise Mine at Boulder, Montana. The discovery of America's first radon health mine was the most rewarding of accomplishments.

Lewis' experiences and knowledge gained during those years culminated in a most important discovery that has benefited multitudes for almost half a century. The epic described by him in *Arthritis and Radioactivity* appears to have been preordained.

John T. Lewis